Bernd Ulmann
Analog and Hybrid Computer Programming

Also of interest

Analog Computing
Bernd Ulmann, 2023
ISBN 978-3-11-078761-0, e-ISBN (PDF) 978-3-11-078774-0

Quantum Information Theory
Concepts and Methods
Joseph M. Renes, 2022
ISBN 978-3-11-057024-3, e-ISBN (PDF) 978-3-11-057025-0

Multi-level Mixed-Integer Optimization
Parametric Programming Approach
Styliani Avraamidou, Efstratios Pistikopoulos, 2022
ISBN 978-3-11-076030-9, e-ISBN (PDF) 978-3-11-076031-6

Automata Theory and Formal Languages
Wladyslaw Homenda und Witold Pedrycz, 2022
ISBN 978-3-11-075227-4, e-ISBN (PDF) 978-3-11-075230-4

High Performance Parallel Runtimes
Design and Implementation
Michael Klemm, Jim Cownie, 2021
ISBN 978-3-11-063268-2, e-ISBN (PDF) 978-3-11-063272-9

Algorithms
Design and Analysis
Sushil C. Dimri, Preeti Malik, Mangey Ram, 2021
ISBN 978-3-11-069341-6, e-ISBN (PDF) 978-3-11-069360-7

Bernd Ulmann

Analog and Hybrid Computer Programming

2nd edition

DE GRUYTER
OLDENBOURG

Mathematics Subject Classification 2010
Primary: 34-04, 35-04; Secondary: 92C45, 92D25, 34C28, 37D45

Author
Prof. Dr. Bernd Ulmann
Schwalbacher Str. 31
65307 Bad Schwalbach
ulmann@analogparadigm.com

ISBN 978-3-11-078759-7
e-ISBN (PDF) 978-3-11-078773-3
e-ISBN (EPUB) 978-3-11-078788-7

Library of Congress Control Number: 2023934962

Bibliographic information published by the Deutsche Nationalbibliothek
The Deutsche Nationalbibliothek lists this publication in the Deutsche Nationalbibliografie;
detailed bibliographic data are available on the Internet at http://dnb.dnb.de.

© 2023 Walter de Gruyter GmbH, Berlin/Boston
Cover image: Bernd Ulmann
Printing and binding: CPI books GmbH, Leck

www.degruyter.com

For Rikka.

"*An analog computer is a thing of beauty and a joy forever.*"[1]

---

[1] JOHN H. MCLEOD, SUZETTE MCLEOD, "The Simulation Council Newsletter", in *Instruments and Automation*, Vol. 31, March 1958, p. 488.

Acknowledgments and disclaimer

This book would not have been possible without the support and help of many people. First of all, I would like to thank my wife RIKKA MITSAM, who never complained about the many hours I spent writing this book. In addition to that, she did a lot of proofreading and post-processed all of the oscilloscope screenshots and various pictures to make them print-ready.

I am also greatly indebted to Dr. CHRIS GILES, who not only gave much constructive criticism but also pointed out lots of additional interesting literature and programming examples. He also extended OCCAM's *razor* into OCCAM's *chainsaw* during the process of proofreading and enhancing this book. :-)

In addition, I wish to express sincere thanks to Dr. DUNCAN CADD, MAIKEL HAJIABADI, FELIX LETKEMANN, BERND JOHANN, NICOLE MATJE, OLIVER BACH, JENS FLEMMER, Dr. CHRISTIAN KAMINSKI, Dr. ROBERT SCHORR and IAN S. KING, who have proofread this book and offered helpful advice. Discussions with JENS BREITENBACH also spotted numerous errors and inconsistencies which were rectified accordingly.

I am also indebted to Mr. MIRKO HOLZER, who programmed the digital portion of the hybrid computer setup described in section 7.5.

Last but not least, I would like to thank TIBOR FLORESTAN PLUTO for his permission to use some of his photographs in this book (figure 6.63 and the title picture).

All of the worked examples in the book have been implemented on an Analog Paradigm Model-1 analog computer for two reasons. First, the author is one of the main developers of this system and second, the machine seems to be the only analog computer currently available on a commercial basis. All of the examples can be (and have been to a large degree) programmed on other machines, like the classic Telefunken or EAI table-top computers if the relatively minor differences in operation and patching are taken into account. Using the Model-1 was not intended for promotional purposes. Many of the examples were previously published online in abbreviated form as application notes.

A wealth of information about typical systems, such as EAI or Telefunken table-top computers, including user manuals, schematics, etc., can be found in the library section of http://analogmuseum.org.

Preface to the 2nd edition

This second edition of "Analog and Hybrid Computer Programming" has been prepared in response to the vastly increased interest in analog computing during recent years. The errors and typos found in the first edition have been corrected and additional topics have been included in the book.

The author is especially indebted to Dr. CHRIS GILES for his invaluable support. He not only did a terrific job proofreading this 2nd edition and was always a great discussion partner when it came to the nitty-gritty details. NICOLE MATJE and OLIVER BACH also did a great job proofreading. STEFAN WOLFRUM also spotted and corrected quite some errors.

The author is also indebted to NICK BABERUXKI and MAIKEL HAJIABADI for the hybrid computing examples shown in chapters 7.6 and 7.7 and their valuable overall feedback.

Since the first edition was published, a new analog computer, *THE ANALOG THING (THAT)*, has been brought to the market as an open hardware project. Consequently, many examples found in this book have been implemented on this small analog computer, which is also described in detail in the introductory chapters.

The following new topics and examples have been included in this second edition:

- Minimum/maximum circuits
- STIELTJES integral
- Transfer functions
- Exponentially mapped past
- SEIR model

- Bessel functions
- The SQ_M model
- Euler spiral
- The Hindmarsh-Rose model of neuronal bursting and spiking
- The simulation of the flight of a glider
- Elastic pendulum
- Making music with analog computers
- Neutron kinetics
- Analog sorting
- Solving systems of linear equations with a hybrid computer approach
- Solving partial differential equations with random walks
- A simple hybrid controller for THE ANALOG THING

Contents

1	**Introduction — 1**	
1.1	What is an analog computer? — 1	
1.2	Direct vs. indirect analogies — 2	
1.3	A short history of analog computing — 3	
1.4	Characteristics of analog computers — 8	
2	**Computing elements — 11**	
2.1	Machine units — 11	
2.2	Summer — 12	
2.3	Integrators — 17	
2.4	Free elements — 24	
2.5	Potentiometers — 25	
2.6	Function generators — 30	
2.7	Multiplication — 33	
2.8	Comparators and switches — 35	
2.9	Input/output devices — 36	
3	**Analog computer operation — 39**	
4	**Basic programming — 47**	
4.1	Radioactive decay — 49	
4.1.1	Analytical solution — 50	
4.1.2	Using an analog computer — 51	
4.1.3	Scaling — 54	
4.2	Harmonic functions — 56	

4.3	Sweep —— 61	
4.4	Mathematical pendulum —— 62	
4.4.1	Straightforward implementation —— 63	
4.4.2	Variants —— 64	
4.5	Mass-spring-damper system —— 65	
4.5.1	Analytical solution —— 66	
4.5.2	Using an analog computer —— 69	
4.5.3	RLC-circuit —— 70	

5	**Special functions —— 73**	
5.1	STIELTJES integral —— 73	
5.2	Inverse functions —— 74	
5.2.1	Square root —— 75	
5.2.2	Division —— 76	
5.3	$f(t) = 1/t$ —— 77	
5.4	Powers and polynomials —— 77	
5.5	Low pass filter —— 78	
5.6	Triangle/square wave generator —— 80	
5.7	Ideal diode —— 80	
5.8	Absolute value —— 82	
5.9	Limiters —— 83	
5.10	Dead zone —— 84	
5.11	Hysteresis —— 85	
5.12	Maximum and minimum —— 85	
5.13	Bang-bang —— 86	
5.14	Minimum/maximum holding circuits —— 87	
5.15	Sample & Hold —— 88	
5.16	Time derivative —— 89	
5.17	Time delay —— 91	
5.17.1	Historic approaches to delay —— 92	
5.17.2	Digitization —— 93	
5.17.3	Sample and hold circuits —— 94	
5.17.4	Analog delay networks —— 96	
5.18	Transfer functions —— 103	
5.19	Exponentially mapped past —— 103	

6	**Examples —— 109**	
6.1	Displaying polynomials —— 109	
6.2	Chemical kinetics —— 110	
6.3	SEIR model —— 114	
6.4	Damped pendulum with external force —— 117	
6.5	MATHIEU's equation —— 119	

6.6	VAN DER POL's equation	**122**
6.7	Generating BESSEL functions	**125**
6.8	Solving the one-dimensional SCHRÖDINGER equation	**127**
6.9	Ballistic trajectory	**130**
6.10	Charged particle in a magnetic field	**131**
6.11	RUTHERFORD-scattering	**135**
6.12	Celestial mechanics	**138**
6.13	Bouncing ball	**141**
6.14	Zombie apocalypse	**144**
6.15	RÖSSLER attractor	**146**
6.16	LORENZ attractor	**148**
6.17	Another LORENZ attractor	**149**
6.18	CHUA attractor	**151**
6.19	Nonlinear chaos	**155**
6.20	AIZAWA attractor	**156**
6.21	NOSÉ-HOOVER oscillator	**157**
6.22	The SQ_M model	**160**
6.23	The DUFFING oscillator	**161**
6.24	Rotating spiral	**163**
6.25	Generating an EULER spiral	**165**
6.26	HINDMARSH-ROSE model	**168**
6.27	Simulating the flight of a glider	**171**
6.28	Flow around an airfoil	**176**
6.29	Heat transfer	**180**
6.30	Two-dimensional heat transfer	**186**
6.31	Systems of linear equations	**187**
6.32	Human-in-the-loop	**193**
6.33	Inverted pendulum	**198**
6.34	Elastic pendulum	**204**
6.35	Double pendulum	**206**
6.36	Making Music	**210**
6.37	Neutron kinetics	**214**
6.38	Smooth sorting	**216**
7	**Hybrid computing**	**219**
7.1	Hybrid controllers	**220**
7.2	Basic operation	**222**
7.3	Shell trajectory	**224**
7.4	Data gathering	**227**
7.5	Training an AI with an analog computer	**230**
7.6	Hybrid solution of systems of linear equations	**236**
7.7	Solving PDEs with random walks	**238**

8	Summary and outlook —— 243	
A	The Laplace transform —— 247	
A.1	Basic functions —— 247	
A.1.1	Step function —— 248	
A.1.2	Delta function —— 249	
A.1.3	Ramp function —— 249	
A.1.4	Exponential and trigonometric functions —— 250	
A.2	LAPLACE transforms of basic operations —— 251	
A.3	Further characteristics —— 252	
A.4	Inverse LAPLACE transform —— 252	
A.5	Example —— 253	
A.6	Block diagrams and transfers functions —— 254	
B	Solving the heat equation with a passive network —— 256	
C	A simple hybrid controller for THE ANALOG THING —— 261	
D	An oscilloscope multiplexer —— 265	
E	A log() function generator —— 269	
F	A sine/cosine generator —— 271	
G	A simple joystick interface —— 273	
H	The Analog Paradigm bus system —— 275	
I	HyCon commands —— 277	

Acronyms

AC	Alternating Current
ADC	Analog to Digital Converter
AI	Artificial Intelligence
BBD	Bucket Brigade Device
DAC	Digital to Analog Converter
DC	Direct Current
DEQ	Differential Equation
DDA	Digital Differential Analyzer
DVM	Digital Voltmeter
EMP	Exponentially Mapped Past
FET	Field Effect Transistor
FPGA	Field Programmable Gate Array
GND	Ground
HPC	High Performance Computing
IC	Initial Condition
LF	Line Feed
NG	Noise Generator
ODE	Ordinary Differential Equation
OP	Operate
PDE	Partial Differential Equation
RAM	Random Access Memory
RC	Resistor Capacitor
RLC	Resistor Inductor Capacitor
SEIR	Susceptible Exposed Infected Recovered
SJ	Summing Junction
THAT	The Analog Thing

1

Introduction

1.1 What is an analog computer?

A book about programming analog and hybrid computers may seem like an anachronism in the 21st century – why should one be written and, even more important, why should you read it? As much as analog computers seem to have been forgotten, they not only have an interesting and illustrious past but also an exciting and promising future in many application areas, such as high performance computing (*HPC* for short), the field of dynamic systems simulation, education and research, artificial intelligence (biological brains operate, in fact, much like analog computers), and, last but not least, as coprocessors for traditional stored-program digital computers, thereby forming *hybrid computers*.

From today's perspective, analog computers are mainly thought of as being museum pieces and their programming paradigm seems archaic at first glance. This impression is as wrong as can be and is mostly caused by the classic *patch field* or *patch panel* onto which programs are patched in form of an intricate maze of wires, resembling real *spaghetti "code"*... On reflection, this form of programming is much easier and more intuitive than the algorithmic approach used for stored-program digital computers (which will be just called *digital computer*s from now on to simplify things). Future implementations of analog computers, especially those intended as coprocessors, will probably feature electronic cross-bar switches instead of a patch field. Programming such machines will resemble the programming of a *field programmable gate array* (*FPGA*), i.e., a compiler will transform a set of problem equations into a suitable setup of the crossbar-switches, thus configuring the analog computer for the problem to be solved.

The notion of an *analog computer* has its roots in the Greek word ἀνάλογον ("*analogon*"), which lives on in terms like "analogy" and "analogue". This aptly characterizes an analog computer and separates it from today's digital computers.

The latter have a fixed internal structure and are controlled by a program stored in some kind of random access memory.

In contrast, an analog computer has no program memory at all and is programmed by actually changing its structure until it forms an analogue, a *model*, of a given problem. This approach is in stark contrast to what is taught in programming classes today (apart from those dealing with FPGAs). Problems are not solved in a step-wise (algorithmic) way but instead by connecting the various computing elements of an analog computer in a suitable manner to form a *circuit* that serves as a model of the problem under investigation. Figures 1.1 and 1.2 illustrate these two fundamentally different approaches to computing. While a classic digital computer works more or less in a sequential fashion, the computing elements of an analog computer work in perfect parallelism with none of the synchronization issues that are often encountered in digital computing.

1.2 Direct vs. indirect analogies

When it comes to analogies in general, it is necessary to distinguish between *direct* and *indirect* analogies, which depend on the underlying principles of the problems being solved and the analogies used to solve them. In short, a direct analogy has its roots basically in the same physical principles as the corresponding problem, i. e., a soap-bubble being used to model a minimal surface, a metal sheet with heaters and thermocouples to investigate heat-flow patterns, etc. If the physical principles underlying the problem and analog computer differ, the computer is called an indirect analog computer.

For the remainder of this book only indirect analogies will be considered, as these are much more versatile in application than their direct counterparts. Typically, such machines are based on analog-electronic computing elements such as summers, integrators, multipliers, and the like.

Although it sounds like a contradiction, analog computers can be implemented using purely digital components. Two such types of machines are *digital differential analyzers* (*DDA*s) and *stochastic computers*, examples of which have been built over many years and whilst they enjoy periodic renaissances, they have never entered the mainstream of computing. Programming these machines follows basically the same lines as programming analog-electronic analog computers (simply called analog computers). These digital analog computers will not be discussed further in the book. More information on DDAs may be found in [FORBES 1957], [FORBES 1972], [WINKLER 1961, p. 215 et seq.], [BECK et al. 1958], [KLEIN et al. 1957, p. 1105 et seq.], [GOLDMAN 1965], [JACKSON 1960, p. 578 et seq.], [ULMANN 2010, p. 157 et seq.], [SHILEIKO 1964], and [BYWATER 1973]. Stochastic computers are covered in [MASSEN 1977].

Fig. 1.1. Principle of operation of a stored-program digital computer (see [TRUITT et al. 1960, p. 1-40])

Fig. 1.2. Structure of an analog computer (see [TRUITT et al. 1960, p. 1-41])

1.3 A short history of analog computing

The idea of analog computing is, of course, much older than today's predominantly algorithmic approach. In fact, the very first machine that might aptly be called an analog computer is the *Antikythera mechanism*, a mechanical marvel that was built around 100 B. C. It has been named after the Greek island Αντικύθηρα (*Antikythera*), where its remains were found in a Roman wreck by sponge divers in 1900. At first neglected, the highly corroded lump of gears aroused the interest of DEREK DE SOLLA PRICE, who summarized his scientific findings as follows:[2]

> "*It is a bit frightening to know that just before the fall of their great civilization the ancient Greeks had come so close to our age, not only in their thought, but also in their scientific technology.*"

Research into this mechanism, which defies all expectations with respect to an ancient computing device, is still ongoing. Its purpose was to calculate sun and moon positions, to predict eclipses and possibly much more. The mechanism consists of more than 30 gears of extraordinary precision yielding a mechanical analogue for the study of celestial mechanics, something that was neither heard of or even thought of for many centuries to come.[3]

---

[2] See [FREETH 2008, p. 7].
[3] See [FREETH 2010].

Slide rules can also be regarded as simple analog computers as they allow the execution of multiplication, division, rooting, etc., by shifting of (mostly) logarithmic scales against each other. Nevertheless, these are rather specialized analog computers, just as *planimeters*, which were (and to some extent still are) used to measure the area of closed figures, a task that frequently occurs in surveying but also in all branches of natural science and engineering.

Things became more interesting in the 19th and early 20th centuries with the development and application of practical mechanical integrators. Based on these developments, WILLIAM THOMSON, later Lord KELVIN, developed the concept of a machine capable of solving differential equations. Although no usable computer evolved from this, his ideas proved very fruitful. Specifically, his approach to programming such machines is still used today and called the *KELVIN feedback technique*.[4]

Figure 1.3 shows a mechanical analog computer, called a *differential analyzer*. On both sides of the long table-like structure in the middle of the picture, various computing elements such as integrators (discernible by the small horizontal disks), differential gears, plotter tables, etc., can be seen. The elongated structure in the middle is the actual interconnect of these computing devices, which consists of a myriad of axles and gears. Programming such a machine was a cumbersome and time consuming process as the interconnection structure had to be more or less completely dismantled and rebuilt every time the machine was configured to solve the differential equations which describe the new problem.

Figure 1.4 shows a simple setup of a differential analyzer to integrate a function given in graphical form. A central motor, shown on the left, drives all computing elements of the machine. On the upper left an *input table* is visible. It consists of a magnifier with crosshairs, which is mounted in such a way that it will be moved by the central motor horizontally while its vertical position is controlled manually by a hand crank, which is turned so that the crosshairs always follow the line of the input function.[5]

At the heart of this setup is an integrator shown at the bottom of the figure. Basically, it consists of a rotating flat disk driven by the central motor and a friction-wheel rolling on the surface of the disk. The radial position of this wheel on the disk is now controlled by the vertical component of the crosshairs on the input table. Given some angular velocity of the rotating disk, the angular velocity

---

[4] Lord KELVIN is often cited as having proposed the use of analog computers for fire control, but although mechanical differential analyzer techniques were successfully employed for purposes such as naval gun fire control in the early 1900s, it took VANNEVAR BUSH to realize that these components could be configured into a general purpose computer.

[5] A steady hand is required for this task, which was quickly automated in order to eliminate this rather unpredictable source of error during a computation.

1.3 A short history of analog computing — 5

Fig. 1.3. VANNEVAR BUSH's mechanical differential analyzer (source: [Meccano 1934, p. 443])

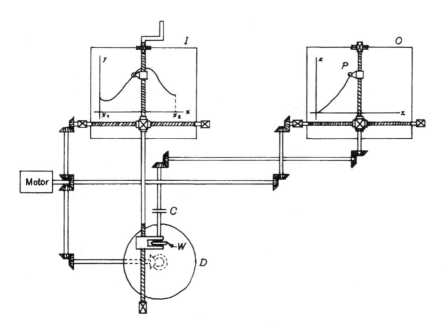

Fig. 1.4. A simple differential analyzer setup for integration (cf. [KARPLUS et al. 1958, p. 190], [SOROKA 1962, p. 8-10])

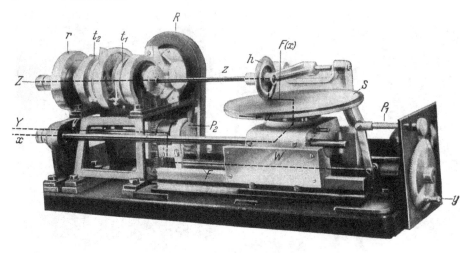

Fig. 1.5. Integrator from the Oslo differential analyzer (see [WILLERS 1943, p. 237])

of the friction-wheel depends on its radial position. If it were located directly above the disk's axis, it would not rotate at all while its angular velocity would be at its maximum if it were positioned at the edge of the disk. Thus, this device effectively performs an integration operation of the basic form

$$\int_0^T f(\tau)\,\mathrm{d}\tau,$$

where τ represents the *machine time* (more about that later) – in this case the rotation of the horizontal disk – and $f(\tau)$ controls the radial position of the friction-wheel. The integrator is running during the time interval $[0, T]$. Figure 1.5 shows an actual implementation of the integrator which was used in the *Oslo differential analyzer*.

The output from the friction-wheel is then used in this setup to control the vertical position of the output table's[6] (upper right of the picture) pen position, while its horizontal position is controlled by the central motor. The resulting figure is the graph of the integral over the input function.

Mechanical differential analyzers like this one were only used for a short period of time as their disadvantages could not easily be overcome. Their setup is cumbersome and time-consuming, the many mechanical parts require a lot of maintenance work, and their speed of computation is limited due to the non-negligible inertias of the rotating and moving parts.

There were some attempts to build electro-mechanical differential analyzers in which basic computing elements were still purely mechanical while their in-

---

[6] Today, this device would be called a *plotter*.

terconnection was accomplished by servo-motors and synchros.[7] The outputs of the synchros could be connected to the inputs of the servo-motors by means of a central electric patch field, thus at least simplifying the basic setup of such a computer.

Mechanical and electro-mechanical analog computers were used in staggering numbers in the form of *fire control systems* during World War II. Being very specialized analog computers, these machines had no direct influence on the further development of the art.

Analog-electronic analog computers were developed independently, beginning in the early 1940s by HELMUT HOELZER in Germany, GEORGE A. PHILBRICK, and A. B. MACNEE in the United States. Their goals were very different.

HOELZER worked in Peenemünde, Germany, on the development of the famous A4 rocket (also known as the *V2*) and was responsible for what would be called its *on-board computer* in today's terms. The result of his work was the world's first electronic stabilization and control system for a rocket. In addition to this, he developed the world's first true general purpose analog computer, which was used during the A4 development. After World War II this unique computer was transferred to the United States and was used at the Redstone Arsenal for further rocket developments well into the 1950s. This machine is shown in figure 1.6.

On the other side of the Atlantic, PHILBRICK's electronic analog computer, named *Polyphemus* due to its peculiar appearance with a single oscilloscope mounted in the top position of its rack, was not aimed at military applications at all. His machine was designed and used to solve problems that typically arise in process control in the chemical industry.

MACNEE's machine, developed at MIT, was a purely academic research instrument and probably the first high-speed electronic analog computer capable of *repetitive operation*, in which a problem is solved over and over again at such high speed that a (nearly) flicker-free picture of the solution can be displayed on an oscilloscope screen.

From today's perspective, these early analog computers seem quite familiar. Except for their particular implementation with vacuum tubes, they already featured all of the typical analog computing elements such as summers, integrators, multipliers, function generators, etc. Furthermore, they were programmed employ-

---

[7] A *synchro* (also known as a *Selsyn*) is basically a rotary transformer with its primary winding on a rotor, which is surrounded by typically three secondary windings. When the primary is fed with an AC signal, signals corresponding to the angular position of the rotor are induced in the stator windings, which can then be used to determine the angle of the rotor. Two synchros can be connected back to back, forming a *Transmitter* and *Control Transformer* pair, which can form the heart of a servo system to perform torque multiplication, one of the biggest challenges in building a mechanical differential analyser. ARNOLD NORDSIECK used these devices in his differential analyzer, see [NORDSIECK 1953] and [BROCK 2019].

ing the same techniques that are still used today. More information on the history of analog computing can be found in [ULMANN 2023][8] and [SMALL 2001].

1.4 Characteristics of analog computers

Although computing by setting up indirect analog computers by myriads of interconnecting wires looks like a disadvantage at first sight, it is probable that this method of programming will soon be replaced by intricate and highly integrated cross-bar switches controlled by an accompanying digital computer. But even a traditional patch panel interconnecting a variety of computing elements has some tremendous advantages over having an algorithm stored in a memory.

First of all, there is no need for memory lookup operations at all in an analog computer, speeding up the overall computation considerably. Further, without any memory there is nothing like a *critical section*, no need to synchronize things, no communications overhead, nothing of the many trifles that haunt traditional parallel digital computers. There is no equivalent to $AMDAHL$'s law[9] in the realm of analog computation. All computing elements work in perfect parallelism.

Another basic advantage of analog computers is their extremely low power consumption, which easily outperforms classic digital computers. This makes analog computing attractive for applications where low power consumption is of utmost importance, such as medical application, embedded devices powered by energy harvesting, etc. Furthermore, analog computers are ideal for high performance computing (HPC) where power is available in abundance and sheer computing power is required.

Finally, analog computers are not prone to problems such as poor stability as is sometimes the case with numerical algorithms for classic digital computers operating on floating point numbers. Even *stiff* differential equations are normally easily solvable by an analog computer whilst many numerical procedures require at least excessive run-times for such problems.

These characteristics of analog computers have led to the recent and impressive increase in interest in this particular approach to computation. The most common form of an analog computer in the near future will be as part of a hybrid computer setup, i.e., closely coupled with a digital computer, thereby relieving it from calculations involving differential equations, etc.

---

[8] German readers might want to refer to [ULMANN 2010] instead.
[9] See [AMDAHL 1967].

Fig. 1.6. HELMUT HOELZER's general purpose analog computer after World War II

2

Computing elements

The following sections introduce the basic elements which comprise an electronic analog computer. Furthermore, the notion of the *machine unit* will be introduced, because the representation of values is of central importance for all of the following concepts. The examples shown have been implemented on an Analog Paradigm Model-1 analog computer.

2.1 Machine units

Voltages or currents are the natural way of representing values within a calculation on an analog computer. Since the majority of historic and modern analog computers use voltages instead of currents, the following sections are restricted to this technique.

Obviously, values represented by voltages are limited by some minimum/maximum voltages, known as *machine units*, m_- and m_+, which are fixed for a given analog computer. Historic vacuum tube based machines often used units of ± 100 V and sometimes ± 50 V, while later and modern analog computers feature machine units of ± 10 V and sometimes even as low as ± 5 V.

All voltages representing variables in a computer setup are bound by these machine units, so it is normally necessary to *scale* a problem to be solved on an analog computer to avoid an *overload* condition in which a variable exceeds the machine unit voltage. If this happens, typically an overload indicator will be lit, identifying the affected computer element. In addition to this, the computer run can be halted automatically to determine the cause of the overload condition. Overloads normally result from erroneous scaling or patching and do not harm the computer but will impair or invalidate the computed results.

Since the machine units are of utmost importance in an analog computer, they are typically highly stabilized with temperature compensated reference elements. With machine units of ± 10 V the computing elements are typically powered by a ± 15 V supply to leave some headroom to detect overloads, etc. So in the case of an overload, the output voltage of the affected element can reach values as high as about ± 15 V on a modern analog computer.

Scaling a problem to be solved on an analog computer has two objectives:

1. Guarantee that no variable exceeds the limits imposed by the machine units.
2. Make the best use of the available interval $[m_-, m_+]$ for each variable of a computer setup to minimize the unavoidable errors caused by the computing elements.

Consequently, it is necessary to distinguish between the *problem variables* in which the problem itself is stated, and the *machine variables* which are the scaled versions of the problem variables. The underlying scaling process is called *variable scaling*. A second scaling process concerns the speed at which the machine will solve a problem in contrast to the speed at which the original problem will act. This is called *time scaling* as it affects the speed of integration and typically does not directly affect the scaling of variables.[10]

Generally, it is a good idea to abstract further from the actual machine units m_- and m_+ and to think within the interval $[-1, 1]$ instead. Analog computers featuring voltmeters as their output devices have those typically scaled accordingly so that the machine units correspond to a display of ± 1 machine units instead of ± 10 or ± 100 Volts.

2.2 Summer

The simplest active element of an electronic analog computer is the *summer*. Its abstract symbol is shown in figure 2.1. A summer yields the *negative* sum of the voltages applied to its inputs at its output, labeled e_o in the figure. Each input has a *weight*, a fixed multiplicative factor applied to the input. Typical weights are 1 and 10, while some machines also feature values of 4 or 5. If no weight is noted next to an input, it is assumed to be 1. Accordingly, all three inputs e_1, e_2, and e_3 in figure 2.1 are weighted by 1.

To understand the behavior of a summer a look at its implementation is necessary. Like most other analog computer elements, it is based on an *operational amplifier*, *opamp* for short, the graphical symbol of which is shown in figure 2.2.

---

10 See section 2.3.

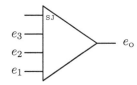

Fig. 2.1. Abstract symbol of a summer with three inputs e_1, e_1, e_2 and *summing junction* input SJ

Fig. 2.2. Graphical symbol of an operational amplifier

An operational amplifier has two inputs, one inverting and one non-inverting, denoted by − and + in the figure. It yields the sum of the values applied to these inputs, amplified by its large (ideally infinite) *open-loop gain* $\mathcal{A}$ that is characteristic for a particular operational amplifier. Typically, gains of 10^5 to 10^9 can be achieved.[11] The use of this type of amplifier in analog computers gave rise to the name *operational amplifier*, as they form the basis of computing elements implementing certain mathematical *operations*.

In a typical analog computer circuit, the non-inverting input of the operational amplifiers is *grounded*, i.e., connected to the analog ground rail, usually denoted by *GND*, which is at the potential representing the value zero; this effectively disables this input.[12]

To build a summer based on an operational amplifier the concept of *negative feedback* is essential. This technique was pioneered by HAROLD STEPHEN BLACK in 1927. The basic idea is to use part of the signal at the output of the operational amplifier and feed it back to its inverting input, thus basically controlling the overall behavior of the resulting circuit by the feedback circuit, instead of relying on the characteristics of the bare amplifier. This idea is central to nearly all operational amplifier circuits, including analog computing elements, such as summers and integrators. Figure 2.3 shows the basic circuit of an operational amplifier with negative (resistive) feedback.

This simple circuit has a single input e_i, which is connected to the inverting input of the operational amplifier by the resistor R_i. The output signal e_o is also connected to the inverting input via a feedback resistor R_f. Since all inputs as well as the feedback path are connected to the inverting input, this is called *summing*

---

[11] Operational amplifiers are rather complex devices. More in-depth information can be found in [JUNG 2006].

[12] In classical analog computers, this non-inverting input was normally not connected to ground directly but used to implement an active drift-compensation. More details on this can be found in [GOLDBERG et al. 1954], [KORN et al. 1964, p. 137 et seq.], [ULMANN 2023, p. 80 et seq.], etc.

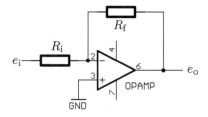

Fig. 2.3. Operational amplifier with negative feedback

junction, SJ or sometimes just S for short. Most implementations of summers (and integrators) make this summing junction available at the patch panel so that additional feedback circuits or additional input resistors, etc., can be connected.

With A denoting the open-loop gain of the operational amplifier (i.e., the gain it exhibits without any negative feedback), and e_{SJ} representing the voltage at the summing junction,

$$e_o = -A e_{\text{SJ}}$$

can be derived for the output voltage of the circuit in figure 2.3.[13] This implies

$$e_{\text{SJ}} = -\frac{e_o}{A} \qquad (2.1)$$

for the voltage at the summing junction itself. Accordingly, the following currents flow into and out of the summing junction:

$$i_i = \frac{e_i}{R_i} \qquad \text{(input current due to } R_i\text{)}$$

$$i_f = \frac{e_o}{R_f} \qquad \text{(feedback current due to } R_f\text{)}$$

$$i_- \approx 10^{-9} \text{ A} \qquad \text{(input current of the amplifier)}$$

Due to KIRCHHOFF's first law, the sum of the currents flowing into and out of a junction must be zero, yielding

$$i_- = i_i + i_f = \frac{e_i - e_{\text{SJ}}}{R_i} + \frac{e_o - e_{\text{SJ}}}{R_f}. \qquad (2.2)$$

Since the input current i_- of a typical operational amplifier is less than a few nA at most, it can be neglected, so that (2.2) simplifies to

$$\frac{e_i - e_{\text{SJ}}}{R_i} = -\frac{e_o - e_{\text{SJ}}}{R_f}.$$

Substituting (2.1) into this yields

$$\frac{e_i + \frac{e_o}{A}}{R_i} = -\frac{e_o + \frac{e_o}{A}}{R_f}.$$

---

[13] All voltages are measured with respect to GND.

Some rearranging results in

$$e_o \left(\frac{1}{\mathcal{A} R_i} + \frac{1}{R_f} + \frac{1}{\mathcal{A} R_f} \right) = -\frac{e_i}{R_i}$$

which can then be solved for e_o:

$$e_o = \frac{-\dfrac{e_i}{R_i}}{\dfrac{R_f + \mathcal{A} R_i + R_i}{\mathcal{A} R_i R_f}} = \frac{-\dfrac{e_i}{R_i} \mathcal{A} R_i R_f}{R_f + \mathcal{A} R_i + R_i} = \frac{-\dfrac{R_f}{R_i} e_i}{1 + \dfrac{1}{\mathcal{A}} \left(\dfrac{R_f}{R_i} + 1 \right)}. \quad (2.3)$$

Since $\mathcal{A}$ is typically very large[14] and R_f/R_i is typically ≤ 10, the denominator of (2.3) can normally be neglected, yielding the simplified form

$$e_o = -\frac{R_f}{R_i} e_i \quad (2.4)$$

describing the output voltage of the feedback circuit shown in figure 2.3.[15]

This shows that the behavior of this circuit is basically determined by the resistors at the input and in the feedback path. If more input resistors are added as shown in figure 2.4, a useful summing circuit results whose overall behavior is readily described by

$$\sum_{i=1}^{n} \frac{e_i}{R_i} = -\frac{e_o}{R_f}. \quad (2.5)$$

The voltage at the output of this circuit is thus the negative of the sum of the voltages at its inputs. The ratios

$$a_i = \frac{R_f}{R_i}$$

define the weights of the various inputs. Typical values for a_i are 1 and 10. If unusual values are required for a certain setup, the necessary resistors can be connected to the summing junction SJ, thus effectively extending the number of inputs of the summer.

---

14 In fact, classical high-precision operational amplifiers used in analog computers had gains of up to $\mathcal{A} = 10^9$.
15 A more informal approach to the behavior of such circuits is to assume that the open-loop gain of the operational amplifier is extremely large, therefore, the voltage at the summing junction is approximately zero. Since the input current of the amplifier is negligible, applying KIRCHHOFF's law to the currents at the summing junction shows that the current through the feedback element is minus the sum of the currents through the input resistors. Applying OHM's law then readily yields the output voltage of the computing element.

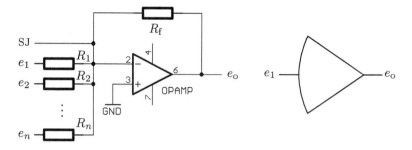

Fig. 2.4. Summer with several inputs based on an operational amplifier with negative feedback

Fig. 2.5. Graphical representation of an open amplifier

Summer basics:

The behavior of an (ideal) summer is described by

$$e_o = -\sum_{i=1}^{n} a_i e_i$$

with the weights a_i being typically 1 or 10. It yields the negative sum of the weighted voltages applied to its inputs. Typical summers have about six inputs, three of which have input weight 1, while the remaining three inputs are weighted by 10.

In some cases it is necessary to disconnect the resistive feedback loop of a summer in order to introduce an external feedback circuit. This case is represented by the symbol shown in figure 2.5. This element is no longer called summer but *open amplifier* or *high gain amplifier* instead. This element always requires some external feedback in order to be stable.

To show the application of summers in a typical analog computer setup, consider the circuit shown in figure 2.6. It solves the equation

$$e_o = -\left(10\left(-\frac{e_1 + e_2}{2}\right) + e_3\right) = 5(e_1 + e_2) - e_3. \tag{2.6}$$

The connection between the output of the first summer and one of its inputs[16] introduces a second feedback resistor parallel to R_f, thus effectively doubling the effect of the feedback loop. This, in turn, halves all other input weights of the summer yielding

$$-\frac{e_1 + e_2}{2}$$

---

[16] Since no explicit weight is denoted next to the input, its weight is equal to 1.

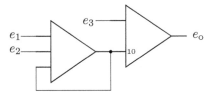

Fig. 2.6. Computer setup according to equation (2.6)

at the output of the left summer. This output is now connected to an input of the right summer, weighted with 10. Another input of this summer is fed with e_3 finally yielding $e_o = 5(e_1 + e_2) - e_3$.

Figure 2.7 shows the front panel of an Analog Paradigm SUM8 This module contains eight summers, each featuring five inputs, three of which have weight 1 and two have weight 10. The four summers in the top row have a special feature which allows the built-in feedback resistor path to be opened by patching a connection between the two jacks labeled FB and $\bot$ (this symbol denotes ground), thus turning a summer into an open amplifier.[17] The summing junction SJ is also available on all eight summers.

2.3 Integrators

At first sight it is astonishing that an analog computer features integration as one of its basic operations. This makes it the ideal machine to solve problems that can be described by differential equations and systems thereof. But there is a catch: Integration on an analog-electronic analog computer is always with respect to time, so all integrations are basically of the form

$$e_o = -\left(\int_0^t \sum_{i=1}^n a_i e_i \, d\tau + e(0) \right) \qquad (2.7)$$

with some constant $e(0)$ called the *initial condition*. In particular, this means that *partial differential equations*, i.e., differential equations in which derivatives with respect to more than one variable occur cannot be handled directly on an analog computer without explicit discretization or other techniques!

---

[17] On these four summers, the feedback resistor R_f is split into a series connection of two resistors of half the size of R_f. The connection between these two resistors is connected to the FB jack. Grounding FB prohibits the current flowing through this path from reaching the summing junction of the operational amplifier, thus effectively disabling the feedback path.

Fig. 2.7. Analog Paradigm SUM8 module

Obviously, the output of an integrator depends on the history of its input values, so this computing element is a kind of memory. In contrast to a simple summer, this necessitates some control of its operation as an integration has a start and an end point in time. It also has to take the initial condition into account. Accordingly, a typical integrator has three modes of operation:

Initial condition: In this mode, often just called *IC* for short, the integrator is *reset* so that its output takes on the negative of the initial condition $e(0)$ that is applied as a voltage to a special input jack of the integrator, which is typically denoted by *IC*. This mode is normally the first step of each computation run on an analog computer. It is important to note that due to time required to charge the capacitor there is a minimum time required by an integrator to take on that initial value.[18]

Operate: The actual computation is done in operate mode, *OP* for short. Here the integrators integrate with respect to time over the negative sum of their inputs as shown in (2.7).

Halt: In some cases it is useful to halt the analog computer for a short period of time (typically of the order of some seconds) to read out values, etc. This is done by switching all of its integrators into HALT-mode, *HALT* for short. In this mode the outputs of all affected integrators are held constant – at least

---

[18] This minimum time depends on the particular setup of the integrator with respect to its *time scale factor* (see below) and the actual analog computer being used.

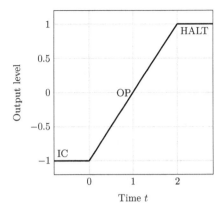
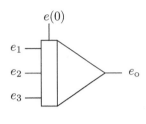

Fig. 2.8. Behavior of a typical integrator **Fig. 2.9.** Graphical representation of an integrator

in theory. Due to technical imperfections some drift is unavoidable, so that the time spent in this mode should be as short as possible to keep errors low. Tolerable HALT-times can be of the order of several tens of seconds.

Figure 2.8 shows the behavior of a typical integrator during these three modes of operation. Here, the IC-input is connected to $+1$ ($+10$ V for example) while one of the integrator inputs weighted with 1 is connected to -1. During IC-mode the output is reset to the negative initial condition, -1. When the OP-mode is activated the actual integration operation with respect to time starts. With this particular setup the integration takes place over a constant value of -1. Accordingly, the output yields a voltage ramp with a slope of 1. After two seconds the analog computer is switched to HALT-mode and the integrator output stays at its last value, $+1$.

The slope depicted in this figure needs some explanations as it is deeply connected to time scaling as mentioned before: A typical integrator features several *time scale factors* k_0 that can be selected by either setting a switch or by placing jumpers between certain plugs on the patch panel. A time scale factor $k_0 = 1$ results in a slope of 1 when integrating over a constant value -1, i.e., with an initial condition of $e(0) = 0$ the integration over -1 will yield an output value of $+1$ after one second in OP-mode. Accordingly, the time scale factor $k_0 = 10$ results in a slope of 10 and so forth. Typical analog computers feature integrators with time scale factors $k_0 = 10^n$ with $n \in \mathbb{N}$ typically ranging from 0 to 3 although some large historic systems, such as the EAI 680,[19] feature time scale factors up to 10^5.

---

19 Short for *Electronic Associates Inc.*

By changing the time scale factors of the integrators in a computer setup it is possible to speed up a calculation or slow it down. This is known as time scaling. Consequently, it is necessary to distinguish between *problem time t*, the time scale the problem under investigation runs in, and *machine time* τ, the time scale the computer operates in, as determined by the time scale factors set on the integrators.

Figure 2.9 shows the graphical representation of an integrator as it is used in analog computer programming. It differs from the symbol for a summer by the addition of a rectangle on its left-hand side, with a vertical bar on its top to which the initial condition value $e(0)$ is connected (this input is always drawn vertically). Like the summer it has a number of inputs with associated weights of typically 1 and 10.

Figure 2.10 shows the front panel of an Analog Paradigm INT4 quad integrator module. Each of its integrators can be set to a time scale factor of 1, 10, 10^2, or 10^3 by a rotary switch and features three inputs weighted by 1 as well as three inputs weighted by 10. In addition to this, each integrator has an initial condition input labeled IC, and a jack yielding $+1$ or -1 in its vicinity as these values are often used as initial conditions.

The two rightmost integrators have two jacks *ModeIC* and *ModeOP*, which allow for direct mode control by feeding appropriate control signals to these jacks. If left unconnected, all integrators are controlled by the central control panel.[20]

The implementation of an integrator is quite similar to a summer with the difference that its feedback path does not consist of a resistor R_f but of a capacitor C instead, as shown in figure 2.11. The feedback current through C is

$$i_f = C\frac{de_o}{dt} = C\dot{e}_o$$

where $\dot{e}_o$ is a short form for

$$\frac{de_o}{dt},$$

a notation that will be used extensively in the following chapters.

Similar to the reasoning in section 2.2 this yields

$$e_o = -\int_0^t a_i e_i \, d\tau \qquad (2.8)$$

with $a_i = C/R_i$ under the assumption of sufficiently large open loop gain $\mathcal{A}$ of the operational amplifier. In fact, a high open-loop gain and small input currents are even more important for a practical integrator than for a summer to keep its error as low as possible, since errors due to the finite gain and to the non-zero amplifier

---

[20] For more information about these control inputs see section 5.15.

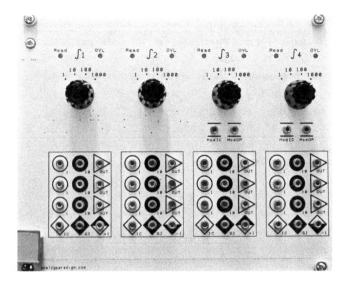

Fig. 2.10. Quad integrator module

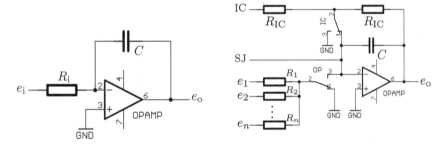

Fig. 2.11. Basic schematic of an integrator

Fig. 2.12. Detailed schematic of an integrator with full mode control

input current naturally accumulate in a circuit like this. As a rule of thumb the condition

$$(1 + \mathcal{A})RC \gg k_0$$

should hold for an integrator.

Equation (2.8) already looks quite similar to (2.7) but still lacks the multiple inputs as well as the initial condition $e(0)$. These are introduced in the more detailed schematic shown in figure 2.12 which can actually be used as a blueprint to build a working integrator. Mode control is implemented by the two (preferably electronic) switches denoted by IC and OP and supports the following three modes of operation:

IC: In IC-mode, the switch labeled OP connects the input resistor array to ground, thus effectively disabling all inputs, while the IC-switch is in the position shown in the figure. Accordingly, the operational amplifier operates basically as a summer with only one input, IC. The voltage over the temporary feedback resistor R_{IC} equals $-IC$ which in turn precharges the capacitor to the negative value of what is present at the IC input jack. The time required to charge the capacitor in this mode of operation now depends on its capacity and the input resistor being active during this mode.

OP: In OP-mode both switches reverse their positions: The IC-switch now grounds the junction between the two resistors labeled R_{IC}, thus disabling the IC-input while the OP-switch connects the input resistor array to the inverting input of the operational amplifier. The capacitor now integrates the weighted sum of the voltages applied to the inputs e_i.

HALT: In this mode, IC is still disabled while the OP-switch ties the input resistor network to ground. Accordingly, the charge of C stays basically constant apart from errors caused by capacitor leakage and the small current flowing into the inverting input of the operational amplifier. These error sources determine the practical length of HALT-times as noted before. HALT mode is typically used to read out values from a computation by means of a digital voltmeter, etc.

It is clearly important for all control switches used by the integrators of an analog computer to switch in perfect synchronism and with no contact bounce at all. Electromechanical relays are only advisable in small, cheap, low-speed analog computers and even there the variation of switching times between different relays can be quite detrimental. Today, electronic switches based on *field effect transistors*, *FET*s for short, are normally used.

Integrator:

The behavior of an integrator is described by

$$e_o = -\left(\int_0^t \sum_{i=1}^n a_i e_i \mathrm{d}\tau + e(0) \right)$$

with the weights a_i being typically 1 or 10 times the time scale factor. The integrator is reset to its initial condition $e(0)$ by activating IC-mode. The actual integration with respect to time takes place during OP-mode. An integration can be temporarily halted by switching the computer to HALT.

The simplest operation cycle of an analog computer consists of the mode-sequence IC-OP-HALT with the time spent in OP-mode either determined manually or by a precision timer. This is typically called *manual control* and is often used on low-speed computations.

Nevertheless, in many cases it is desirable to display a more or less flicker-free image of variables vs. time on an oscilloscope display and this requires an automated mode of operation called *repetitive operation*. Here a precision timer cycles the integrators repetitively through the mode-sequence IC-OP-IC-OP-... With sufficiently high time scale factors it is possible to repeat a full simulation run every couple of milliseconds, thus making it possible to display a set of graphs on an oscilloscope.

Since it is, of course, possible to change the coefficients and initial conditions of a computer setup manually at every instant, repetitive operation can give an immediate insight into the behavior of a problem with respect to its parameterization as one can see the effects of parameter changes instantaneously on the display.

A variant of this mode of operation is called *single run* and consists of a single sequence IC-OP-HALT with preset IC- and OP-times. This is especially useful when using a pen plotter as output device where a single run will draw a particular solution of a problem.

The computer setup shown in figure 2.13 illustrates repetitive operation. It consists of two integrators in series where the leftmost integrator is fed with the constant $+1$. A summer with only one input is used to invert the sign of the output of this first integrator.[21]

Setting both integrators to the same time scale factor $k_0 = 10^n$ and setting the OP-time to $2 \cdot 10^{3-n}$ milliseconds, e. g., $k_0 = 100$ and OP-time 20 ms, yields a linear voltage ramp ranging from $+1$ to -1 at the output of the leftmost integrator. This signal is then fed into two inputs with weight 1 of the second integrator. These two paralleled inputs result in a combined input weight of 2. Since the IC-input of this integrator is connected to -1, too, its output voltage starts at $+1$ at the beginning of the OP-period and follows a parabola afterwards.

This setup is described mathematically by $y_1 = 1$ and the voltage ramp

$$y_2 = -\left(\int_0^t y_1 \, d\tau - 1\right) = 1 - t \text{ with } 0 \leq t \leq 2$$

at the output of the leftmost integrator which is fed into the second integrator finally yielding

$$y_3 = -\left(2\int_0^t (1-\tau)d\tau - 1\right) = 1 - 2t + t^2 = (t-1)^2.$$

---

21 Such a summer is often called an *inverter*.

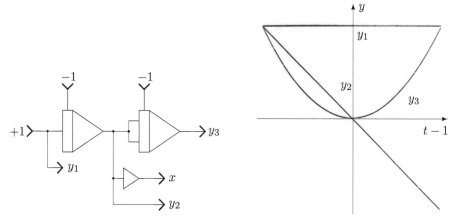

Fig. 2.13. Generating a parabola by double integration over a constant

Fig. 2.14. Parabola generated by integrating twice over a constant function

Figure 2.14 shows an oscilloscope screenshot with the analog computer running in repetitive mode containing all three functions y_1, y_2, and y_3.[22]

2.4 Free elements

Using the summing junction input of summers and integrators, *free elements* such as resistors, diodes, and capacitors may be employed to extend the input capabilities of these computing elements. In addition to that they can be used to establish specialized feedback paths in order to setup computer circuits yielding functions such as absolute value, hysteresis, signum function, etc., as will be shown later.

Figure 2.15 shows the front panel of three typical Analog Paradigm free element modules. The module on the far left is called *XIR* as it contains two free resistor networks which can be connected to the summing junction of any summer or integrator in order to extend the number of available inputs. In addition to the standard input with weights of 1 and 10, this module also features input resistors with weights 0.1 and 100.

The module in the middle, *XID*, contains six SCHOTTKY diodes as well as two 10 V ZENER-diodes, sometimes called *Z-diodes*, which are often used to implement limiters, etc.

---

[22] The actual display is, of course, bright on a dark background which has been postprocessed for printing here.

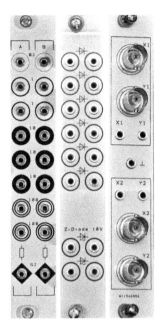
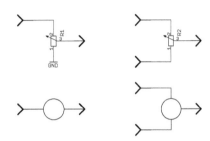

Fig. 2.15. Typical free elements such as resistors, diodes, and BNC adapters

Fig. 2.16. Basic circuits used as coefficient potentiometers

The module on the right, *XIBNC*, holds four BNC jacks which can be used to connect peripheral equipment such as oscilloscopes, function generators, etc., to the analog computer system.

2.5 Potentiometers

A machine without the possibility of setting coefficients in a setup would be pretty useless as soon as problems get even slightly more complicated than the simple parabola shown above. In analog-electronic analog computers *coefficient potentiometers* (also called just *potentiometers* or colloquially "*pots*") are used as adjustable attenuators to introduce coefficients. The simplest form of such a device is just a potentiometer used as a *voltage divider*.

Figure 2.16 shows the two basic circuits for coefficient potentiometers at the top as well as the graphical representation used in analog computer programming. A simple voltage divider as shown on the left allows setting coefficients ranging from 0 to 1. In some cases a *free potentiometer* as shown on the right half of the figure is employed to continuously cross-fade between two input values.

A voltage divider basically can be seen as consisting of an upper resistor R_u and a lower resistor R_ℓ connected in series between the input signal and ground.

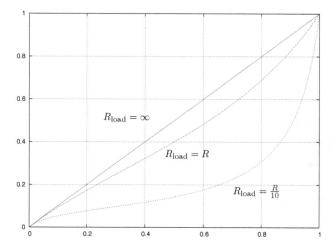

Fig. 2.17. Voltage divider with loads $R_{\text{load}} = \infty$, $R_{\text{load}} = R$ and $R_{\text{load}} = R/10$

The output is connected to the connection between both resistors. The proportion between the two resistors is determined by the angular position α of the potentiometer dial:

$$R_u = (1 - \alpha)R$$
$$R_\ell = \alpha R$$

Obviously, $\alpha = 0$ yields no output signal at all since the output connection is then effectively tied to ground.

Since the overall resistance of this setup is $R_u + R_\ell = R$, the output voltage e_o will be

$$\frac{e_o}{e_i} = \frac{R_\ell}{R}$$

in the *unloaded* case for an input voltage e_i. In this context unloaded means that no current flows out of the voltage divider, thus the same current flows through both resistors. In this unusual yet idealized case the coefficient can be directly set to up to three decimal places by means of a precision ten-turn potentiometer fitted with a multi-turn dial.

Since most computing elements typically have a finite input resistance between $10^4\ \Omega$ and $10^6\ \Omega$, the output of such a voltage divider is *loaded* resulting in a non-negligible setup error as shown in figure 2.17. The linear solid line shows the behavior of an unloaded potentiometer with respect to α, while the dashed and dotted lines show the effect of external loading with R and $R/10$ respectively.

A loaded voltage divider is described by

$$R_\ell \parallel R_{\text{load}} = \frac{R_\ell R_{\text{load}}}{R_\ell + R_{\text{load}}}$$

with $\|$ denoting a parallel circuit. Accordingly, the overall resistance of this setup is

$$R_{\text{total}} = R_u + R_\ell \| R_{\text{load}} = R_u + \frac{R_\ell R_{\text{load}}}{R_\ell + R_{\text{load}}}$$

yielding the following expression for the output voltage of a voltage divider loaded by R_{load} for a given setting α:

$$\frac{e_o}{e_i} = \frac{R_\ell \| R_{\text{load}}}{R_{\text{total}}} = \frac{\frac{R_\ell R_{\text{load}}}{R_\ell + R_{\text{load}}}}{R_u + \frac{R_\ell R_{\text{load}}}{R_\ell + R_{\text{load}}}} = \frac{R_\ell R_{\text{load}}}{R_u(R_\ell + R_{\text{load}}) + R_\ell R_{\text{load}}}$$

$$= \frac{R_\ell}{\frac{R_u R_\ell}{R_{\text{load}}} + R_u + R_\ell} = \frac{\alpha R}{(1-\alpha)R \frac{\alpha R}{R_{\text{load}}} + R} = \frac{\alpha}{(1-\alpha)\alpha \frac{R}{R_{\text{load}}} + 1}.$$

This makes potentiometer setting a rather tedious process on most analog computers as the value shown on the precision dial typically deviates quite substantially from the actual value with the deviation depending on the actual computer setup, i.e., the load connected to the slider of the potentiometer.

Therefore, most analog computers feature an additional mode of operation called *potentiometer setting*, *POTSET* for short, in which the integrators are switched to IC- or HALT-mode and all input resistor networks are disconnected from the inverting inputs of their associated operational amplifiers and connected to ground instead. Thus, all inputs show the correct load resistance to their feeding computing elements without any computation being performed.

In this mode the input of a potentiometer selected to be setup is automatically connected to +1. Its output value – with its correct load – can then be read out by means of a high-impedance *digital voltmeter*, or *DVM* for short. Instead of a DVM a *compensation voltmeter* is often used by many early or small analog computers. Here, the output voltage of the potentiometer to be setup is compared against a precision voltage source. The difference between these two values is then fed to a null-detector. With both voltages being equal, no current flows between the potentiometer being setup and the reference source, thus no additional error is introduced. Nevertheless, this method is quite time-consuming as one has to setup the precision voltage source first and then adjust the actual coefficient potentiometer until the null-detector shows that the adjustments match.

With the advent of affordable operational amplifiers it became possible to employ a *impedance converter* in order to unload the potentiometer, as shown in figure 2.18. The only load at the potentiometer slider is then due to the negligible current i_+. The potentiometer is said to be *buffered* in this case.

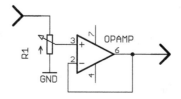

Fig. 2.18. Potentiometer with impedance converter

The clever circuit extending this idea shown in figure 2.19 makes it possible to set a coefficient $-1 \leq a \leq +1$ instead of restricting its value to the interval $[0, 1]$ as in a simple voltage divider. The behavior of this circuit is described by

$$e_o = e_1(2\alpha - 1)$$

with $0 \leq \alpha \leq 1$.

Figure 2.20 shows an example of the application of a coefficient potentiometer in an analog computer program for computing the integral

$$e_o = -\int_0^t 5e_i \, d\tau = -5\int_0^t e_i \, d\tau.$$

It should be noted that the classic graphical symbol for a potentiometer does not distinguish between its input and output, since this will always be clear from the surrounding circuitry.[23] Some classic analog computer programs rely on the fact that the potentiometers used are unbuffered. Such programs may have to be adapted when being ported to a machine featuring only buffered potentiometers.

Figure 2.21 shows a typical PT8 module from an Analog Paradigm analog computer containing eight buffered potentiometers, the last of which can be used as a free potentiometer. While the first seven potentiometers each feature one input and two paralleled outputs, this eighth potentiometer has two input jacks labeled INa and INb. If it is to be used as a standard coefficient potentiometer, a connection between INb and $\perp$ must be patched. All of these potentiometers are of the ten-turn type and are fitted with precision dials, allowing them to be set up to a maximum of three decimal places. Due to their associated buffers, the dial readings are accurate.

---

[23] Connecting a potentiometer erroneously may result in its destruction or at least blow a tiny fuse which protects the potentiometer against overloads in classic analog computers. On modern machines with impedance converters following the voltage dividers there is typically no risk of damaging something by a patching error.

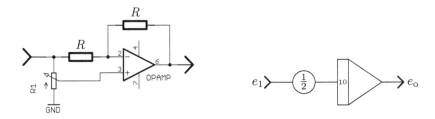

Fig. 2.19. Extended coefficient circuit allowing for factors $-1 \leq a \leq 1$

Fig. 2.20. Using a coefficient potentiometer

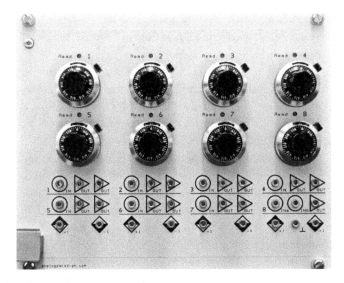

Fig. 2.21. Potentiometer module

Potentiometers:

Potentiometers always have a distinct input and output, which must not be reversed although their graphical symbol in computer programs won't distinguish explicitly between input and output. In classic analog computers the sliders of the precision ten-turn potentiometers are typically protected by either incandescent light bulbs or by microfuses. Be careful not to blow these microfuses as these are hard to get nowadays. More recent implementations usually employ an operational amplifier as impedance converter to unload the potentiometer's slider and protect it from patching errors. Typically potentiometers can only be used to set coefficients $0 \leq a \leq 1$. Some designs give a parameter range of $-1 \leq a \leq 1$, but these normally require a DVM for setup as no suitable precision dials are available.

2.6 Function generators

Sometimes it is necessary to generate arbitrary functions as part of an analog computer setup to solve a complex problem. It is always advantageous if such functions can be generated by solving some auxiliary function or by using an abbreviated TAYLOR *series* to give a local approximation. Unfortunately, this is sometimes not possible, for example when functions are based on experimental data, etc. In these cases, *(arbitrary) function generators* are employed. Figure 2.22 shows the graphical symbol of such a device.

Although there is a wide variety of approaches to generating arbitrary functions,[24] the following section is limited to the most widely used approach: the polygonal approximation. Generally, arbitrary functions of more than one variable are not easily implemented on an analog computer and are often subject to a high degree of approximation errors.

Figure 2.23 shows a simplified schematic of a typical polygon function generator. The idea is to use biased diodes that start conducting at different values of the input variable e_i. In this example there are three diodes, which start conducting at voltages determined by the settings of the potentiometers R1, R2, and R3. The respective currents they contribute to the summing junction of the operational amplifier are determined by the potentiometers R4, R5, and R6 respectively.

It is assumed in the following that the voltages at which the diodes start to conduct are set so that the topmost diode starts conducting first, followed by the diode in the middle, and finally the bottom one while e_i runs from -1 to $+1$. With only the topmost diode conducting, the output of the operational amplifier is a voltage depending linearly on e_i, its slope being controlled by the setting of R4. As soon as the second diode starts conducting, the operational amplifier will yield the (negative) sum of both input currents contributed by these two diodes depending on e_i, etc.

The potentiometers R1, R2, and R3 thus control the *break-points* while R4, R5, and R6 control the slope of each of the linear functions to be added by the operational amplifier yielding a polygonal line approximating the desired function.

The diodes acting as electronic switches are not perfect diodes, i.e., they do not start conducting immediately at the desired break-point but instead exhibit

---

[24] Cf. [KORN et al. 1964, p. 207 et seq.].

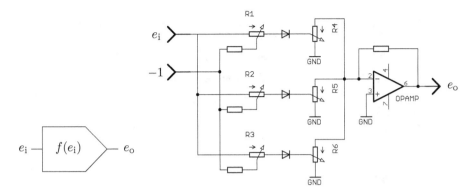

Fig. 2.22. Graphical representation of a function generator

Fig. 2.23. Simplified schematic of a function generator with biased diodes

a parabolic behavior for values in a small interval about the break-point; this is normally not a problem. In addition to this, the kinks of a polygonal line are sometimes undesirable. These can be smoothed out by superimposing a small high-frequency sinusoidal signal on e_i, if required.

Figure 2.24 shows a typical example of such a function generator (of course having many more diodes than shown in the schematic of figure 2.23) made in the 1960s by EAI. It features ten diode segments, each with adjustable break-point and slope. The potentiometers on the left side control the break-points, while those on the right set the slope of each segment. The rotary switches in the middle control the basic direction of the slope (positive, negative, or off). Such an arrangement has the advantage that the break-points can be located according to the "complexity" of the function being approximated.

Another widespread implementation variant of such diode function generators features equally spaced fixed break-points. While EAI always used variable break-points, the German company *Telefunken* mainly used function generators with fixed break-points arguing that this makes their setup simpler since only the slope potentiometers have to be set. Giving up the freedom of choosing break-points freely is, indeed, not too much of a loss as most functions are relatively well-behaved, not requiring excessive slopes for a polygonal approximation. The device shown in figure 2.25 contains four distinct function generators, each having 21 slope potentiometers.

Some functions are used frequently enough to justify the construction of fixed function generators. Among these functions are sin(), cos(), tan(), exp(), and log(). Figure 2.26 shows a module from a Telefunken analog computer of the 1960s featuring eight such special functions.

Setting up such diode function generators is quite time-consuming and sometimes better approximations or even functions of more than one variable, which

Fig. 2.25. Telefunken function generator with fixed break-points

Fig. 2.24. Diode function generator with adjustable break-points (EAI)

Fig. 2.26. Example of a classic fixed function generator (Telefunken)

cannot be realized with reasonable effort by purely analog electronic means, are required. In cases like these, a digital computer can be coupled to the analog computer by means of analog-digital- and digital-analog-converters.[25] The digital computer may then perform simple and fast table-lookups based on the values read from its ADC(s) and output the corresponding function values by means of a DAC. This, in fact, is a first step towards a hybrid computer.

Function generators:

Most function generators are based on using biased diodes to approximate functions by polygonal lines. If possible, it is advisable to create special functions by other means such as TAYLOR approximation, etc. Function generators are something of a last resort due to their time-consuming setup and the errors introduced by approximating typically smooth functions by polygons.

---

25 *ADC*s and *DAC*s for short.

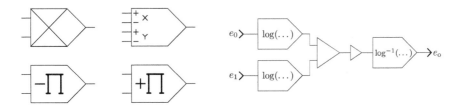

Fig. 2.27. Graphical symbols for multipliers

Fig. 2.28. Multiplication by adding logarithms and taking the antilog

2.7 Multiplication

Interestingly, multiplication is much more complicated to implement with analog-electronic circuits than integration. Multiplication of two variables $x \cdot y$ can be seen as a function of x and y and is thus not easily amenable to the simple function generator approach as described above. Over the past 50 years many ingenious multiplication circuits have been devised,[26] but today the GILBERT cell multiplier is most often used as it offers high precision and can be easily implemented in an *integrated circuit (IC)*.

Figure 2.27 shows the most commonly used graphical symbols for multipliers. Modern multipliers only need two inputs, x and y, to compute $x \cdot y$, which applies to the symbol in the upper left and the two symbols in the second row. If a certain multiplier performs a sign-inversion, the symbol in the lower left is used to emphasize this behavior. The symbol in the upper right is included for mostly historic reasons, as it applies to the *quarter-square multiplier*, which requires both variables x and y with positive and negative signs each.

One of the most straightforward approaches to implementing a multiplier is the use of two logarithm- and one antilog-functions as shown in figure 2.28. This is analogous to the technique on which slides rules are based. The input values must obviously satisfy $e_0, e_1 > 0 + \varepsilon$ with ε determined by the log function generators used to avoid an overload. Furthermore, special precautions are necessary to deal with negative signs of the arguments. Under these constraints, this circuit can also be used to implement division by subtracting one logarithmic value from the other.

Another, historically more widely employed, implementation of a multiplier made use of the fact that

$$xy = \frac{1}{4}\left((x+y)^2 - (x-y)^2\right).$$

---

[26] Cf. [KORN et al. 1964, p. 254 et seq.] for more information.

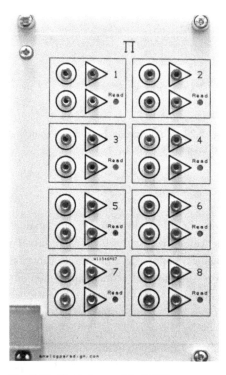

Fig. 2.29. Multiplier module MLT8

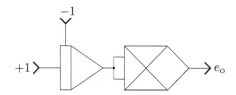

Fig. 2.30. Generating a parabola with an integrator and a multiplier

This requires only two function generators instead of the three required for the logarithm-technique described above. Likewise, the arguments are no longer restricted to a positive interval. Due to its use of two squaring functions in conjunction with the factor of $1/4$, this type of multiplier became known as quarter square multiplier. Its implementation typically required positive and negative versions of both of its arguments e_0 and e_1. Often these values were readily available in the surrounding computer setup. If not, dedicated inverters (summers with only one input) were used.

Multipliers allowing arguments $-1 \leq e_0 \leq +1$ and $-1 \leq e_1 \leq +1$ are generally called *four quadrant multipliers* in contrast to *two quadrant multipliers*, which only yield results in two of the four quadrants of a Cartesian coordinate system.

Figure 2.29 shows a modern Analog Paradigm MLT8 multiplier module containing eight individual four quadrant GILBERT-cell multipliers.

A simple application is shown in Figure 2.30 where a multiplier in conjunction with an integrator is used to generate a parabola. The integrator yields a linear voltage ramp running from $+1$ to -1 which is fed into both inputs of a multiplier, which in turn yields a parabola function at its output.

Multiplication, etc.:
Many traditional analog computers feature quarter square multipliers which need both input variables with positive and negative signs. By reversing the two inputs for either x or y a sign-reversal of the result can be achieved without the need for an additional inverter.

2.8 Comparators and switches

A *comparator* is used to compare two analog signals e_0 and e_1 and to yield a logic output signal for the condition $e_0 + e_1 > 0$.[27] Typically, a relay or preferably an electronic switch is controlled by this logic output signal. This makes it possible to implement switching functions in an analog computer setup.

Figures 2.31 and 2.32 show the graphical symbols used to denote a comparator-switch combination as well as a pure comparator. In contrast to a relay switch, which has no dedicated inputs and outputs, an electronic switch may not be reversed with respect to its input/output connections.

Figure 2.33 shows a simplified schematic of a typical comparator driving a relay. An operational amplifier is fed with the two analog inputs to be compared via two resistors R of equal value. Its feedback path consists of two biased diodes which start conducting at certain output voltages of the operational amplifier as determined by the two bias sources, which are depicted here as batteries. The purpose of these two diodes is to limit the output voltage in order to avoid driving the output stage of the amplifier into saturation. Classic implementations often controlled high-speed polarized telegraph relays with this circuit.

Figure 2.34 shows an Analog Paradigm CMP4 module with four comparators with four associated electronic switches. The left half of each row belongs to the comparators with their two inputs and the output while the right half is associated with the electronic *SPDT*-switches. The output of each comparator as well as the control input of each associated electronic switch are available at the front panel. If the input jack of an electronic switch is left unconnected, it is automatically connected to its associated comparator's output. If the output of a comparator is explicitly patched to the input of an electronic switch, it will take over control of this switch. This makes it possible to control more than one switch with a single comparator output without requiring additional connections for normal operation.

---

[27] It should be noted that a typical comparator does not perform a comparison like $e_0 > e_1$.

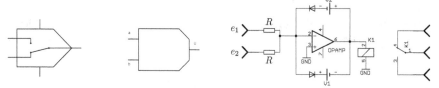

Fig. 2.31. Graphical symbol representing a comparator combined with a switch

Fig. 2.32. Symbol for a comparator yielding a logic output signal

Fig. 2.33. Basic circuit of a comparator with relay

Comparators and switches:

Comparators employing electromechanical relays as their switching parts should be used with care and not in conjunction with high-speed computations as the time-lag and the time variation in closing and opening the switches of different relays triggered at the same time can introduce substantial errors into a computation. Wherever possible electronic switches should be used in conjunction with comparators.

The relays used in conjunction with the comparators in many classic analog computers are quite fragile and can be damaged by patching errors such as shorting m_- to m_+ or to ground by its contacts. Electronic switches typically have dedicated inputs and outputs, in contrast to electromechanical relays.

2.9 Input/output devices

Analog-electronic analog computers are ideally suited to be interfaced directly with sensors and actuators that operate with analog signals. Consequently, there is a plethora of possible input/output devices that can be employed with analog computers. The two most common output devices are the pen-plotter and the oscilloscope. While the pen-plotter is normally used with the analog computer set to single run mode, allowing one trace of a function to be plotted in x, y-mode, the oscilloscope is typically used in conjunction with either repetitive operation or with the analog computer being in OP-mode for an extended period of time. This makes it possible to create a more or less flicker-free picture on the oscilloscope screen, which, together with the possibility of changing any coefficient during a computer run manually, often gives great insight into the behavior of the problem being investigated.

Figures 2.35 and 2.36 show typical outputs generated by a pen-plotter and an oscilloscope.

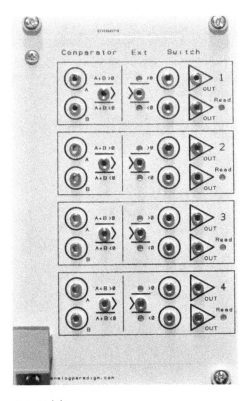

Fig. 2.34. Quad comparator module

Fig. 2.35. Example of a pen-plotter output

Fig. 2.36. Typical oscilloscope screenshot

3
Analog computer operation

Although, at first sight, using an analog computer to solve a problem may appear to be a daunting prospect, in practice it is a relatively straightforward process. Figures 3.1 and 3.2 show two classic analog computers – an EAI-580 and a Telefunken RA 770 precision analog computer.

The most obvious feature of both machines is the large, central patch panel, which contains several thousand jack sockets; these are typically shielded on high precision analog computers. Whilst most manufacturers developed their own cable and connector system, small table-top analog computer use standard 4 mm or 2 mm banana plugs for interconnecting their computing elements. The analog computer program is set up by linking the jacks on the patch panel with the patch leads to create a *circuit* which is an analogue for the problem being investigated.

Figure 3.3 shows a patch panel being removed from such a classic machine. Programs were typically pre-patched and the patch panels stored in a cabinet when not in use. As archaic as this seems from today's perspective, it is a viable means of storing and "loading" analog programs.

A more detailed look at the EAI-580 in figure 3.1 shows that its left half is divided into four subunits. At the bottom is the main control panel allowing the operator to select the desired mode of operation (IC, OP, HALT, repetitive operation). It also contains the push-button controls for setting up the servo-set potentiometers, which were normally fitted in medium to large analog computers. These replaced most of the manual precision potentiometers and consist of a potentiometer mechanically coupled to a small motor that is in turn controlled by a simple servo-loop. Using such an arrangement it is possible to set up these potentiometers by selecting a potentiometer address and entering the desired value on this control panel. Without servo-set potentiometers the number of manual potentiometers and variable function generators, which had to be set up, made program changes time-consuming and cumbersome.

40 — 3 Analog computer operation

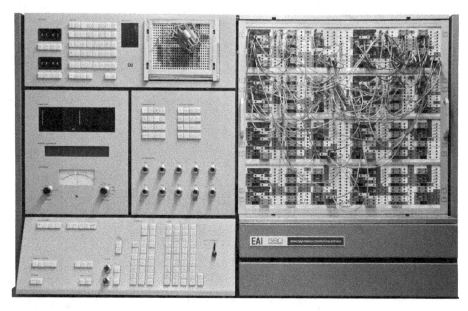

Fig. 3.1. EAI-580 analog computer

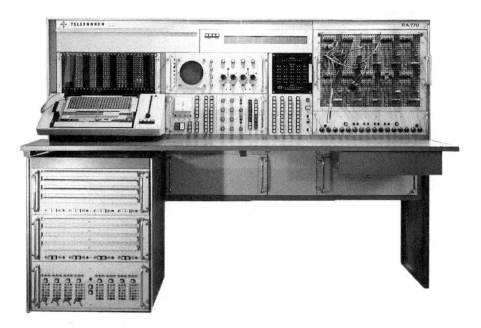

Fig. 3.2. Telefunken RA 770 precision analog computer

Fig. 3.3. Removing a patch panel from an EAI 231RV analog computer (cf. [EAI PACE 231R, p. 6])

The leftmost panel in the middle section contains the central overload indicator panel, which features one incandescent light bulb for every operational amplifier in the system, as well as a four-digit digital voltmeter and a traditional moving coil instrument. The rightmost panel contains ten manual potentiometers and a number of switches and display lights, which show the respective states of the computer's comparators and relays.

On top is a digital expansion unit that consists of a number of logic gates, flip-flops, and delay elements, which can be interconnected by means of a small dedicated removable patch panel. Using these digital elements, it is possible to extend the simple repetitive mode of operation by controlling the modes of individual integrators, etc. This facility is useful for solving complex problems such as process optimization.

The Telefunken RA 770 computer shown in figure 3.2 contains basically the same elements. The operator controls and an oscilloscope are located in the middle section of the computer. On the left side is a large digital expansion group, similar to that of the EAI-580 system. The expansion chassis on the bottom left contains variable and fixed function generators.

An example of a modern analog computer, the Analog Paradigm Model-1, is shown in figure 3.4. Compared with classic machines it is tiny and does not have a central patch panel due to the fact that the overall system is fully modular and thus can be equipped with the modules that any given problem requires. The

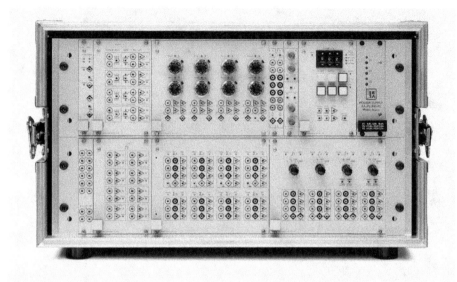

Fig. 3.4. Analog Paradigm Model-1 analog computer (picture by JAMES BALL)

system shown is the most basic configuration consisting of eight summers, four integrators with four time scale factors each, eight multipliers, four comparators with electronic switches, eight manual precision potentiometers, a power supply, and a control unit. Since modern analog computers are not normally shared between different users, the absence of a central patch panel is not a significant drawback and is more than compensated for by the inherent flexibility of the modular system design.

The control unit, CU for short, of this system is shown in more detail in figure 3.5. It controls the operation of the integrators in the computer either in manual or automatic mode. In manual mode the computer can be switched into the modes IC, OP, and HALT by means of the push-button INITIAL and the toggle-button OP/HALT.

Automatic mode allows either single or repetitive operation of the analog computer by activating the SINGLE or REPEAT button respectively. In both of these cases, OP-time is set in 10 ms steps by the push-button precision potentiometer on the top of the CU. The push-button ICTIME selects one of two preset IC-times, suitable for integrators operating with a time constant of 1 or 10 and 10^2 or 10^3 respectively. Pushing OVLOAD will cause the computer to automatically enter its HALT-state when an overload is detected in any computing element during an OP-cycle.

The jacks on the bottom of the control unit provide trigger signals that can be used in conjunction with external devices such as oscilloscopes or plotters.

Fig. 3.5. Control unit of a Model-1 analog computer

The jack labeled EXTHALT can be used to halt the current OP-cycle by applying an external signal, typically the output of a comparator or a signal derived from external equipment.

An even more recent analog computer is *THE ANALOG THING*[28] (*THAT*) which was introduced as an open hardware project[29] in late 2021 and is shown in figure 3.6. This system cannot compete with large classic analog computers with respect to precision or its sheer number of computing elements, but it is powerful enough to implement really interesting analog computer programs.

It features eight coefficient potentiometers, five integrators[30], four summers, four inverters, two multipliers, two comparators, two sets of free resistor networks, which can be used to extend the number of inputs of the integrators, summers, and inverters, as well as several free capacitors and diodes. When larger problems are to be tackled, several THATs can be cascaded in a master/minion mode of operation by means of connectors on the back.

---

28 See https://the-analog-thing.org, retrieved December 1st, 2022.
29 The complete hardware documentation of THE ANALOG THING can be found on github: https://github.com/anabrid/the-analog-thing, retrieved on December 1st, 2022.
30 Each of these integrators features two time scaling factors of $k_0 = 10$ and $k_0 = 10^3$ with the latter being the default setting. If $k_0 = 10$ is required, OUT of this integrator must be connected with its SLOW jack.

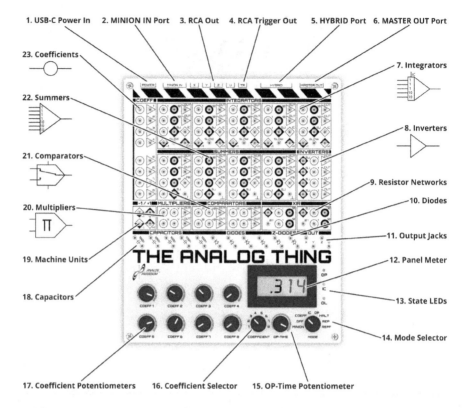

Fig. 3.6. THE ANALOG THING (see [FISCHER 2022, p. 8])

Integrators, summers, and inverters have their summing junctions accessible through jacks labeled SJ. This can be used to connect to an external resistor network such as the two denoted by XIR on the left, thus adding further inputs to a computing element. Connecting a diode or Z-diode between the output of a computing element and its summing junction implements a limiter function.

On the back of the system a variety of connectors can be found. On the left is the power connector (a USB-C jack), following by MINION IN, which allows the computer to be part of a chain of THATs. The RCA-jacks[31] labeled X, Y, Z, and U are connected to the corresponding jacks on the front panel by means of 1 : 10 voltage dividers. This makes it possible to use a standard sound card to gather data from this analog computer. The TR jack yields a trigger signal, which can be used to start data gathering by a data logger or oscilloscope. The next connector, HYBRID, can be used to control the operation of THE ANALOG THING by an

---

[31] These are often called phono jacks.

attached digital computer, while MASTER OUT connects to MINION IN of the next system in a master/minion setup.

Operation of THE ANALOG THING is straightforward. The switch on the lower right controls the mode of operation. When set to MINION the mode of operation of this THAT is controlled by the master in a chain, OFF turns the device off. Coefficients are set in COEFF mode. Since no precision 10-turn potentiometers are used in this little analog computer, the coefficient to be set is selected by the switch labeled COEFFICIENT. Its corresponding value is then displayed on the panel voltmeter.

IC sets the machine to initial condition, while OP and HALT select operating and halt mode respectively. The two switch positions labeled REP and REPF allow repetitive operation with the operation time set by the OP-TIME potentiometer. REP allows for operation times of up to several seconds, while REPF is used for fast repetition times. In both cases, the panel voltmeter gives an indication of the selected operational time.

4
Basic programming

Although solving a problem on an analog computer has basically the same prerequisites as programming a digital computer, the actual process of programming differs significantly from the now-predominant algorithmic approach. Figure 4.1 shows the main steps involved. First, the problem under investigation has to be examined and a mathematical representation, typically a system of coupled differential equations, *DEQ*s for short, has to be derived. From here on the digital and analog ways of programming quickly diverge. On a digital computer these equations form the basis for an algorithm which will be executed in a stepwise fashion. When programming an analog computer the very same equations will be transformed into a patching setup which determines how the various computing elements are connected to each other. This setup is called an *analog computer program*.

Since differential equations are of utmost importance in analog computer programming, a few words about these mathematical objects might be appropriate. Basically, a differential equation is an equation in which the unknown is not a simple value but rather a *function*. A differential equation is made up from this unknown function as well as various derivatives of this function. To simplify things a bit[32] it will be assumed that the function and all of its required derivatives are with respect to a single independent variable. This independent variable is often

---

[32] Differential equations are by no means really simple – many of them, often those of great relevance for practical applications, have no closed form solution at all and are thus only accessible to numerical or – in our case preferably – analog simulation techniques.

48 —— 4 Basic programming

Fig. 4.1. How to program an analog computer (cf. [TRUITT et al. 1960, p. 1-108])

time. Differential equations of this type are commonly called *ordinary differential equations* or *ODEs* and are of the general form

$$f\left(t, y, \frac{dy}{dt}, \frac{d^2y}{dt^2}, \ldots, \frac{d^ny}{dt^n}\right) = 0 \quad (4.1)$$

where y denotes the unknown function, not a "simple" variable! If the independent variable is time t, y depends on t and thus should be always thought of as $y(t)$, although this is almost never written explicitly to avoid unnecessary clutter in the equations. This time dependency makes differential equations very interesting and powerful, as they can be used to describe dynamic systems.

The highest derivative of y determines the *order* of the ODE. As is customary, derivatives with respect to time will be denoted by simply placing a dot over the function as shown in the following:

$$\dot{y} = \frac{dy}{dt}, \quad \ddot{y} = \frac{d^2y}{dt^2}, \quad \dddot{y} = \frac{d^3y}{dt^3}, \ldots$$

Typically, physical systems with y representing some position of an object require derivatives $\dot{y}$ (*velocity*), $\ddot{y}$ (*acceleration*), and sometimes $\dddot{y}$ (*jerk* or *jolt*).

Differential equations can also contain functions and derivatives thereof with more than one independent variable. A typical example for this is the well known one-dimensional *wave equation*

$$\frac{\partial^2 u}{\partial t^2} = c_p^2 \frac{\partial^2 u}{\partial x^2},$$

which describes how a wave travels through a medium. Differential equations of this type are called *partial differential equations (PDEs)* and are normally much more complicated to solve, even by using an analog computer, than ODEs.

4.1 Radioactive decay

As an introductory example, the behavior of radioactive decay will be investigated in some detail using an analog computer. Most chemical elements appear in different *isotopes*, which have the same number of protons in the nucleus but differ in their respective numbers of neutrons. Not all of these isotopes are stable. Some are unstable and these typically emit particles such as an α- or β-particle (a Helium nucleus and an electron, respectively) or electromagnetic γ-radiation. These are the most common, but by no means the only ways in which unstable isotopes can transmute into other isotopes through *radioactive decay*.

Regardless of the actual mechanism of the decay, the number of atoms decaying in a given time interval is always proportional to the amount of the isotope at the start of the interval. As radioactive decay is a stochastic process, the focus will typically be on large numbers of atoms of any given isotope. With the dot notation as introduced before, this can be written as[33]

$$\dot{N} \propto N,$$

where $\propto$ denotes proportionality which can be expressed by a multiplicative factor λ, the *decay rate*, as follows:

$$\dot{N} = -\lambda N. \tag{4.2}$$

This is a typical, yet very simple, ODE describing how an unstable isotope decays over time. As before, N is a function of time. The obvious question now is which function will satisfy the ODE (4.2) in general?

4.1.1 Analytical solution

This equation is simple enough to allow an analytical solution, which will be derived below. ODEs simple enough to be solved by such analytical means are quite rare, but fortunately the existence of such a solution is not necessary for programming an analog computer.[34] However, the analytical solution derived here can be used later to verify the solutions obtained by the analog computer.[35]

Rearranging (4.2) yields

$$\frac{\dot{N}}{N} = -\lambda.$$

Integrating with respect to t on both sides gives

$$\int \frac{\dot{N}}{N}\,dt = -\int \lambda\,dt = -\lambda t + c$$

with c denoting the constant of integration. The left side of this equation equals $\ln(N)$[36] yielding

$$\ln(N) = \lambda t + c.$$

Applying exp() on both sides and explicitly writing the argument t this results in

$$N(t) = e^{-\lambda t + c} = e^{-\lambda t} e^c.$$

---

[33] In this book, time as an argument to a function is typically not explicitly shown, so N and $\dot{N}$ have to be read as $N(t)$ and $\dot{N}(t)$ respectively.

[34] Nevertheless, appendix A gives a short overview of the methods of the *operational calculus*, which can be used to solve differential equations analytically under certain circumstances.

[35] The approach used here lacks the rigor required from a mathematician's point of view.

[36] This can be shown by explicit differentiation:

$$\frac{d}{dt}\left(\ln(N) + c\right) = \frac{d}{dt}\ln(N) = \frac{1}{N}\frac{d}{dt}N = \frac{\dot{N}}{N}$$

With $N(0)$ denoting the number of atoms existing at $t = 0$ we get
$$N(0) = e^0 e^c = e^c,$$
finally yielding
$$N(t) = N(0)e^{-\lambda t}, \qquad (4.3)$$
the *universal law of radioactive decay*[37] with λ denoting the *decay rate*.

Typically, the *half-life* $T_{1/2}$ is found in isotope tables instead of λ. $T_{1/2}$ denotes the time in which half of the atoms available at $t = 0$ have decayed. It is linked to λ by
$$T_{1/2} = \frac{\ln(2)}{\lambda},$$
yielding the following form for the exponential decay that is typically found in textbooks:
$$N(t) = N(0)e^{-\frac{\ln(2)}{T_{1/2}}t}.$$

4.1.2 Using an analog computer

As previously noted, it is unusual that a differential equation describing a "real" problem is so simple that it can be easily solved analytically. In cases where no analytical solution is possible or if the work required to do so is too great, an analog computer can be used to simulate the behavior of the dynamic system described by the DEQ under investigation. To accomplish this the equation or the system of equations have to be transformed into a computer setup or program as will be shown in the subsequent sections. This can be done in two main ways, using either the KELVIN *feedback technique* or the *substitution technique*.[38]

In the following example, the KELVIN feedback technique, which is quite straightforward, will be used. Here, the differential equation is first rearranged so that its highest derivative is isolated on the left side of the equation. It is then assumed that this derivative is known. Feeding it to a chain of integrators, with summers where necessary to invert signs, etc., all lower derivatives on the right-hand side of the equation will be obtained. These lower derivatives are then combined using suitable computing elements to yield this highest derivative from which the successive integrations started. This output is then connected to the input of the integrator chain, thus forming a *feedback loop*.

---

[37] Another derivation, lacking the mathematical rigor, can be found in standard texts such as [SCHWANKNER 1980, p. 111 f.].
[38] Cf. [SCHWARZ 1971, p. 168 et seq.] for more information.

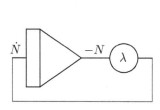
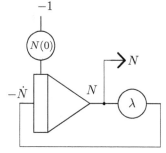

Fig. 4.2. Initial computer setup for the decay problem

Fig. 4.3. Final computer setup for the decay problem

In the case of radioactive decay the DEQ already has its highest derivative on the left-hand side:

$$\dot{N} = -\lambda N. \tag{4.4}$$

Pretending that $\dot{N}$ is known, it can be fed to an integrator which readily yields $-N$ at its output.[39] Multiplying this value by λ results in $-\lambda N$, which is the right-hand side of (4.4). Since the left and right sides have to be equal, $-\lambda N$ can now be used as the input for the integrator used to derive $-N$, effectively forming the central feedback loop shown in figure 4.2.

This setup is not yet complete as it lacks a way to set the initial condition $N(0)$. It is also desirable to have N as output value instead of $-N$, which can be easily achieved by changing the signs of the left and right side of equation (4.4). The program shown in figure 4.3 satisfies both of these requirements.

The resulting setup of these two subcircuits on THE ANALOG THING is shown in figure 4.4. Shown is the upper left corner of the patch panel. The input of the first coefficient potentiometer is connected to -1 and yields the desired initial condition $N(0)$ at its output which is in turn connected to the IC input of the first integrator. One of the two outputs of this integrator is connected to the second coefficient potentiometer, the output of which is connected to an integrator input with weight 1. Using this potentiometer λ can be set. The second output[40] is connected to the X jack on the lower right of the patch panel (not shown here). The corresponding output at the rear of THE ANALOG THING is connected to an oscilloscope.

A typical solution curve can be seen in the oscilloscope screenshot shown in figure 4.5. The solution displayed is clearly of the form $\exp(-x)$ as expected from the analytical solution (4.3).

---

[39] The implicit change of sign caused by integrators and summers must always be accounted for.

[40] All output jacks of these computing elements are connected to each other.

Fig. 4.4. Setup for the simulation of radioactive decay

Fig. 4.5. Simulation of radioactive decay

In order to obtain deeper insight into the behavior of a dynamic system like this, the computer can be run in repetitive mode. If the OP-time is short enough, a (nearly) flicker-free picture of solution curves can be obtained on an oscilloscope screen. The effects of changing the settings of the coefficient potentiometers will immediately show on the display. This mode of operation typically requires a simple linear ramp τ for the x-deflection of the oscilloscope.[41] This can be easily generated by a single integrator, as shown in figure 4.6. The setting of β controls the position at which the oscilloscope beam will start, while α controls the speed at which the ramp increases, thus controlling the width of the resulting picture.

Figure 4.7 shows a long-term exposure screenshot with the analog computer running in repetitive mode using a linear voltage ramp for x-deflection, while the parameter λ was varied manually during the exposure. The exponential character of the decay function can be seen clearly.[42]

While a high time scale factor k_0 of the integrators is required for an oscilloscope output, k_0 must be much lower if a plotter is used as output device in order to not exceed the maximum writing speed of the plotter. In this case single run operation would be used to obtain a single graph of the solution function per computer run.

---

[41] It would also be possible to use the built-in sweep generator of a standard oscilloscope, but this would require triggering the oscilloscope from the analog computer's control unit in order to obtain a stable picture.

[42] The faint trace running from the lower right to the upper left is the result of the periodic reset of the integrators. Ideally the oscilloscope's beam should be switched off by its z-input under control of the control unit.

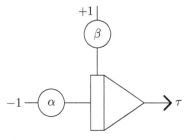
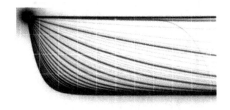

Fig. 4.6. Generation of a sweep signal

Fig. 4.7. Simulation of radioactive decay in repetitive mode of operation with varying parameter λ

4.1.3 Scaling

The solutions obtained in the previous section were only qualitative – the settings of λ and k_0 were chosen arbitrarily to obtain a feeling for the behavior of the exponential decay. To get quantitative solutions the problem must be scaled properly. Scaling is a central part of any analog computer programming and is often the most difficult aspect.

As a rule of thumb, all variables within a computation should span as much of the permissible range of values $[-1, 1]$ as possible in order to keep the effects of unavoidable errors in the computing elements as small as possible. Therefore, $N(0)$ might be scaled so that it corresponds to -1.[43] If, for example, $0.12 \; mols$[44] of $^{234}_{90}\text{Th}$ are given, this would be represented by a setting of $N(0) = 1$ in the computer setup shown in figure 4.3. If the output N of the inverter reads 0.4 after some time, this would correspond to $0.4 \cdot 0.12 = 0.048$ mols.

So, it is necessary to distinguish between *problem variables*, which are the variables appearing in the mathematical representation of the problem, and *machine variables*, which are the corresponding scaled variables as they occur in a given scaled computer setup. Machine variable are often denoted by a hat:

$$\widehat{N} = \alpha N \text{ with } \alpha = \frac{1}{\max(N)},$$

---

[43] It does not matter what the actual voltage of the machine unit is – be it 5, 10, 50, 100 V or something completely different. As long as everything is scaled to ± 1 the actual machine unit only has to be taken into account during readout if a standard DVM or other instrument is used.

[44] One mol of a substance contains about $6.022 \cdot 10^{23}$ atoms or molecules (AVOGADRO's constant) and weighs as many grams as the atomic/molecular weight of that substance is. So one mol of $^{12}_{6}\text{C}$ weighs 12 grams.

where α is the scale factor for this variable. In the example above it is

$$\alpha = \frac{1}{0.12} \approx 8.333.$$

This amplitude scaling of variables is only one side of the coin. It is also necessary to distinguish between *problem time*, i.e., the time in which the actual problem runs, and *machine time*, the time at which the current simulation runs. The process of transforming problem time to machine time is called *time scaling*. To distinguish between both times the symbols t and τ are commonly used, representing problem and machine time; these are coupled by a scale factor β yielding $\tau = \beta t$. β is determined as the product of k_0 of the integrators in a given setup and their respective input weights.

Machine time is running faster than problem time if $\beta > 1$. This is suitable for simulating processes which change too slowly with time for direct observation. Radioactive decay of some long-lived isotope is a typical example of such a process. In other cases where the problem time is too short to make useful observations, choosing $\beta < 1$ suitably will slow down machine time accordingly.

In the case of $^{234}_{90}\text{Th}$ having a half-life of $T_{1/2} = 24.1$ days, a substantial speedup of machine time as compared to problem time is desirable. To get a flicker-free display on an oscilloscope, machine time will be scaled so that 2.41 ms will correspond to the half-life $T_{1/2} = 24.1$ d with $0 \leq \tau \leq 20$ ms. This results in

$$\lambda = \frac{\ln(2)}{2.41} \approx 0.287,$$

which will be set using the coefficient potentiometer labeled λ in figure 4.3. Setting the time scale factor of the integrator to $k_0 = 10^3$ will scale everything from seconds to milliseconds, yielding the desired time-compression.

Scaling:
Scaling often isn't as easy as in the example above. The following approach should be generally used:
1. Determine the interval for each variable in a computer setup. This can be done by either
   - guestimating (possible only if the underlying differential equations are not too complicated) or
   - by solving the problem numerically, or
   - experimentally, by running the analog computer and observing the variables on an oscilloscope, etc. If a variable causes an overload, it is a good idea to scale it down by a constant factor such as 2. If a variable is contained in a small subinterval of $[-1, 1]$, it should be scaled up accordingly to minimize errors due to the implementation of the computing elements.

2. Variables to be scaled are typically the outputs of a summer or integrator. In case of division, both enumerator and denominator have to be taken into account during the scaling process. To scale a variable x by a factor ξ the following steps are necessary ($\xi < 1$ scales down while $\xi > 1$ scales up):
   - Scale all inputs to the summer or integrator yielding x by ξ.
   - In all places where x is used as input to following summers or integrators, a scale factor of $1/\xi$ has to be introduced. This is not directly possible with multipliers, dividers, etc., as these computing elements do not feature inputs with weights > 1. In these cases the inverse scaling factor must be taken into account at a later stage.
3. These steps have to be performed for all variables in a given problem. Eventually, most of the scaling factors introduced in the steps above will cancel out yielding the final computer setup.

There also exists software, such as the DEQscaler,[45] to perform problem scaling for analog computers. This package basically solves the equations in question numerically, determines maximum values and then derives suitable scale factors for an analog computer implementation.

4.2 Harmonic functions

The next example is a bit more complicated and is based on the following second order differential equation

$$\ddot{y} + \omega^2 y = 0. \tag{4.5}$$

In order to apply KELVIN's feedback technique equation (4.5) has to be rearranged so that its highest derivative appears on the left side yielding

$$\ddot{y} = -\omega^2 y.$$

This differential equation is still simple enough that analytic solutions can be "guessed". Functions y for which the second derivative is equal to the negative of the function with some weight ω^2 are required. One such function is, of course,

---

[45] See https://github.com/bernd-ulmann/DEQscaler, retrieved December 14th, 2022. This page also contains an introductory example showing the application of the automatic scaler.

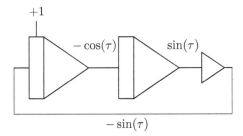

Fig. 4.8. Basic computer setup for equation (4.5) with $\omega = 1$

the constant 0-function, which is quite trivial. Two more interesting solutions are based on sine and cosine as

$$\frac{d^2 \sin(\omega t)}{dt^2} = -\omega^2 \sin(\omega t) \text{ and } \frac{d^2 \cos(\omega t)}{dt^2} = -\omega^2 \cos(\omega t).$$

So just by inspection, three basic solutions of this differential equation have been identified. These are called *particular* solutions, but which solution is the correct one? This cannot be decided from just (4.5) as this equation does not contain any information about the initial conditions. Since this DEQ is of order two there are two initial conditions involved, one for y, called $y(0)$, and one for $\dot{y}$, denoted by $\dot{y}(0)$.

To keep things simple assume that $\omega = 1$. This results in the basic computer setup shown in figure 4.8. The leftmost integrator has an output signal of -1 at the start of the simulation run, $t = 0$. The second integrator implicitly has an initial condition of 0, therefore, its output is 0 at $t = 0$ and will initially rise during the simulation due to the input values delivered by the output of the first integrator.

With $\omega = 1$, τ is controlled solely by k_0, which has to be set equal for both integrators. Setting $k_0 = 10^3$ results in an output signal with a period of 6.28 ms, which is exactly what would be expected for $\omega = 1$.

Using a different initial condition scheme such as 0 for the leftmost integrator and -1 for the second integrator would have yielded $\cos(\tau)$ instead of $\sin(\tau)$. By applying initial conditions $\neq 0$ to both integrators, linear combinations of these two particular solutions can be achieved.

Since this basic circuit yields $\pm \cos(\tau)$ as well as $\pm \sin(\tau)$ for suitable initial conditions, it is commonly called a *quadrature generator*. It is the basis of many more complex computer setups and is especially useful when figures based on closed loops need to be displayed on an oscilloscope, as will be shown in later sections.

Now, what about $0 < \omega < 1$? Wouldn't a simple solution just require a single coefficient potentiometer after the inverter in the setup shown in figure 4.8? Unfortunately, this is not the case as ω effectively affects the time scaling of both integrators! Just using one potentiometer following the inverter, thus preceding the input of the leftmost integrator, would just change its time scale factor to

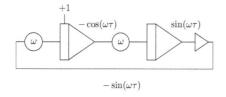

Fig. 4.9. Basic computer setup for equation (4.5) with variable ω

Fig. 4.10. Output of the second integrator in the computer setup shown in figure 4.9 with $k_0 = 10^3$ and $\omega = \frac{1}{2}$

ωk_0 while the second integrator would still run at k_0. Therefore, it is typically necessary to change the time scale factor of both integrators at once, as shown in figure 4.9.

Figure 4.10 shows the resulting output signal of the leftmost integrator for $\omega = \frac{1}{2}$. One x-division of the screen corresponds to 2 ms as before. So one period of the displayed sine-signal is about 12.55 ms, which is a good approximation to 4π as would be expected for this particular value for ω.

The results shown so far have been generated with the computer set to repetitive operation and a OP-interval of 20 ms. In some cases a circuit like that shown in figure 4.9 has to run for an extended period of time, for example to display a ball (a scaled-down unit circle based on a sine/cosine signal pair) on an oscilloscope as it bounces around in a box. In such cases a simple computer setup like this normally won't suffice as the amplitude of the resulting sine/cosine signal pair will not be constant over extended periods of time. This results from the unavoidable imperfections of real computing elements. Depending on the computer used the amplitude may decrease or increase over time, either collapsing to zero or causing an overload condition of the integrators and the inverter.

Therefore, some additional circuitry is required for *amplitude stabilization*. The idea behind this is fairly simple. On one hand, to avoid a decreasing amplitude, some positive feedback must be introduced. This then makes a limiter circuit necessary to avoid the ever-increasing amplitude that would otherwise inevitably result. These two measures will, of course, cause some, but normally negligible, distortion of the resulting waveform, especially when the sine/cosine signals are used to drive some sort of display.

Figure 4.11 shows a typical setup of a quadrature generator with amplitude stabilization: At its heart is the circuit shown before, with the difference that the central feedback loop uses integrator inputs weighted with 10, so that the overall time scale factor is $10 \omega k_0$. The potentiometer labeled α introduces some positive feedback into the loop thus steadily increasing the amplitude of the output signal. Depending on the analog computer used this positive feedback may be unnecessary. If the amplitude decreases without it, it is sufficient to set α to a very small value (less than 10^{-2}, typically). To avoid running into an overload

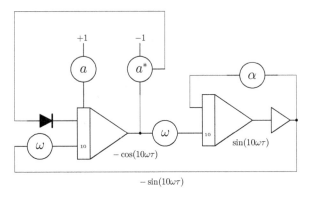

Fig. 4.11. Basic computer setup for equation (4.5)

some negative feedback is necessary to keep the overall amplitude stabilized. This is implemented by the potentiometer a^*, which is one of the rare cases where a free potentiometer is required. The output (wiper) of that potentiometer is connected to a diode that is in turn connected to an integrator input weighted with 1.

The idea here is to apply a negative bias to the diode so that it starts conducting only when the output of the leftmost integrator is higher than this bias voltage. Then the diode will start conducting, essentially implementing a negative feedback loop (with negative exponential characteristics). It is important that the weight of the input connected to the diode is at least a factor 10 smaller than the input for the main feedback loop.

This setup works well on analog computers with buffered coefficient potentiometers, such as the Analog Paradigm Model-1 used here. If a classic analog computer without buffered potentiometers is used, the cathode of the diode connected to the potentiometer a^* should be connected to the summing junction of the leftmost integrator.[46]

There is some interplay between the settings of the potentiometers labeled a and a^*. As a determines the initial amplitude of the output signal, a^* should be set so that the diode bias voltage is of the same size as this initial amplitude. If $a > a^*$, the amplitude stabilization circuit will kick in directly after setting the computer to OP-mode. This will result in a quickly decreasing amplitude which is normally undesirable.

---

[46] An even simpler implementation saves the potentiometer a^* and connects the output of the leftmost integrator to its summing junction by means of two ZENER-diodes in anti-series. These diodes will start conducting when the amplitude reaches their combined threshold level, thus effectively limiting the output amplitude of the integrator.

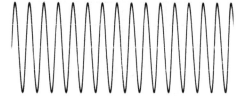

Fig. 4.12. Output of the second integrator in the computer setup shown in figure 4.11 with $k_0 = 10^3$ and $\omega = \frac{1}{2}$ resulting in a time scale factor of $10 k_0 \omega = 5000$

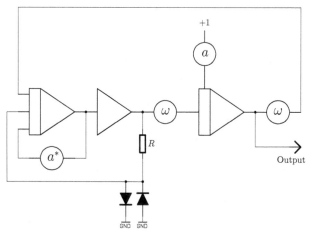

Fig. 4.13. A different approach to an amplitude stabilized oscillator

Figure 4.12 shows the output of this circuit with the analog computer set to continuous operation with $\omega = \frac{1}{2}$ and $k_0 = 10^3$. The oscilloscope's x-deflection is set to 2 ms per division as before.

A variation of this circuit, which is shown in figure 4.13, has been devised by EDWARD L. KORTE[47]. At first sight, this circuit looks quite similar to that described above, but there are two subtle differences:

- The amplitude is stabilized by a negative linear feedback on the leftmost integrator as controlled by the setting of the potentiometer labeled a^*.
- It requires some positive feedback to keep the amplitude up. This is implemented by the resistor R and two limiting diodes. The resistor should be of the same order as the input resistors of the leftmost integrator. This results in a small positive rectangular feedback signal which will keep the amplitude from falling to zero.

---

[47] See [KORTE 1964].

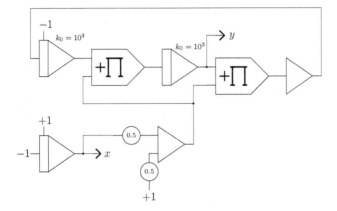

Fig. 4.14. Basic sweep circuit

This circuit is extremely useful when a high-frequency harmonic signal or signal pair with constant frequency is required in a computation. Nevertheless, it is not well-suited as the basis for a sweep generator, as described in the following section, since the amplitude setting a^* is quite frequency dependent.

4.3 Sweep

Sometimes it is useful to have a harmonic function *sweep* over a frequency range. In a laboratory environment this is typically done with a *sweep generator*, also known as *wobbulator*. Solving the differential equation

$$\ddot{y} = -\omega^2 y$$

with ω varying linearly from 0 to 1 readily yields a harmonic sweep function. This is a particular useful variation of the computer setup shown in the preceding section. The basic setup is shown in figure 4.14.

The underlying idea is simple. The bottom integrator generates a voltage running linearly from -1 to $+1$. Scaling this down by a factor of 0.5 and adding 0.5 yields a linear ramp running from 0 to -1 at the output of the summer connected to the output of this integrator. The two potentiometers used in figure 4.9 to set ω have been replaced by two multipliers in this circuit, which are fed with this voltage ramp as the time-varying value for ω.

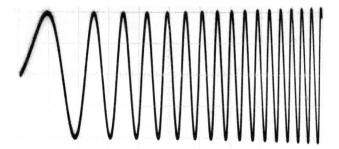

Fig. 4.15. Waveform generated by solving $\ddot{y} = -\omega^2 y$ with $0 \leq \omega \leq 1$

The resulting output signal is shown in figure 4.15. It should be noted that the amplitude could use some stabilization, as shown before, because the amplitude visibly increases with time.[48]

Basically, the subcircuit in the upper half of figure 4.14 can be used to generate $\sin(\varphi)$ and $\cos(\varphi)$ if $\dot{\varphi}$ is given and fed to the two multipliers. This is actually a very useful circuit since many problems require harmonic functions for a wide range of the parameter φ and have $\dot{\varphi}$ available elsewhere in the setup, as in the following example.

4.4 Mathematical pendulum

This section describes the simulation of the movement of the tip of a *mathematical pendulum* as shown in figure 4.16. It consists of a weightless rod which is pivoted at point 0 around which it can swing without friction. Mounted at the other end of the rod is a punctiform mass m. The angle α_0 denotes the maximum displacement of the pendulum, i.e., its initial condition, while α represents the current angle with respect to the vertical running through the pivot located at 0.[49]

The mass is subjected to the force $F_g = mg$ which is caused by the acceleration due to gravity. This can be resolved into in two forces acting tangentially (F_t) and radially (F_r) on the mass:

$$F_t = -mg\sin(\alpha)$$
$$F_r = mg\cos(\alpha)$$

---

[48] The output signal should be taken from the integrator following the one with the amplitude stabilization circuitry. This will yield a less distorted signal due to the roll-off effect of the integration.

[49] Since α changes with time, $\alpha(t)$ would be more accurate, but it would clutter the following derivations. Furthermore, the time-dependency of this angle will show in the dot-notation used for its derivatives.

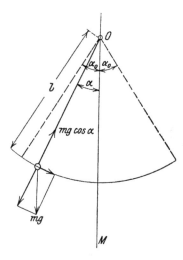

Fig. 4.16. Mathematical pendulum (cf. [HORT 1910, p. 1])

F_t is the restoring force, which will drive the mass back to its position of rest. It causes a tangential acceleration

$$a_t = \frac{F_t}{m} = l\ddot{\alpha}$$

with l denoting the length of the pendulum rod and $\ddot{\alpha}$ being the angular acceleration. This readily yields

$$ml\ddot{\alpha} = -mg\sin(\alpha)$$

and thus the following differential equation of second order describing the dynamic behavior of the pendulum:

$$\ddot{\alpha} + \frac{g}{l}\sin(\alpha) = 0 \tag{4.6}$$

4.4.1 Straightforward implementation

If it wasn't for the term $\sin(\alpha)$ in this equation, it could be easily solved by an analog computer using only summers, integrators, and coefficient potentiometers. This sine term makes things a little bit more complicated. If a sine function generator is available, it can be used to mechanize this equation. Otherwise one has to resort to one of the following two approaches:

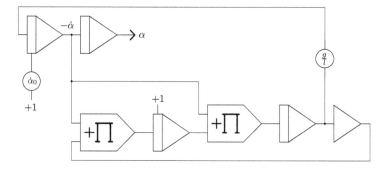

Fig. 4.17. Pendulum simulation with sine generator circuit

If α is sufficiently small,[50] $\sin(\alpha)$ can be approximated as

$$\sin(\alpha) \approx \alpha,$$

which simplifies (4.6) to

$$\ddot{\alpha} = -\frac{g}{l}\alpha,$$

which is just (4.5) with $\omega^2 = g/l$, so the pendulum can be simulated with the computer setup already shown above in figure 4.9.

If α is not sufficiently small, either a sine function generator is required or one can resort to using the sweep circuit described above to generate the sine function based on $\dot{\alpha}$ as its argument instead of α. Fortunately, $\dot{\alpha}$ is already available as $\ddot{\alpha}$ has to be integrated anyway. Figure 4.17 shows the resulting computer setup with a sine generator controlled by $\dot{\alpha}$.

A word of caution: As useful as such a sine/cosine generator subcircuit is, it is subject to drift and to multiplier balance errors. Generally, it should not be used for extended simulation runs and is best employed with the computer set to repetitive operation with short OP-times.

4.4.2 Variants

Although this implementation is quite elegant, it comes at a cost as it requires two multipliers, two integrators, and one summer just for the generation of the sine function. In cases like this mathematical pendulum, where the argument of the trigonometric function is restricted (which is the case as long as the pendulum does not flip over), other implementation variants may be advantageous.

---

[50] What *sufficiently small* means actually depends on the precision required for a solution of this problem. Typically, $-\pi/8 \leq \alpha \leq \pi/8$ can be considered small enough as the error will stay below 1%.

A straightforward approach would be to use the TAYLOR series

$$\sin(\varphi) = \sum_{n=0}^{\infty} (-1)^n \frac{\varphi^{2n+1}}{(2n+1)!}$$

to approximate the required harmonic function. Depending on the desired accuracy, which in turn depends on the range of the argument φ, it will typically suffice to set

$$\sin(\varphi) \approx \varphi - \frac{\varphi^3}{6} \qquad (4.7)$$

for arguments restricted to $-\pi/4 \leq \varphi \leq \pi/4$.

This function can be readily generated by two multipliers, a summer, and a coefficient potentiometer, thus not only saving some additional computing elements but also eliminating drift problems that might accumulate over time using the method described in the preceding section.

Certain powers of time-dependent variables can be easily generated using two integrators with appropriate time scale factors in series without requiring any multipliers at all. In this case one could start from

$$\int \left(\int \varphi \, d\tau \right) d\tau = \frac{1}{6} \varphi^3$$

yielding the second term of (4.7). Changing the sign of this solution and adding φ will then yield the required sine approximation, an approach that will be detailed in section 5.4.

4.5 Mass-spring-damper system

Apart from the amplitude stabilization scheme shown in section 4.2, the problems shown so far have exhibited oscillating behavior with no damping. The next example introduces damping using the simple mass-spring-damper system shown in figure 4.18. In this example y denotes the vertical position of the mass with respect to its position of rest. Neglecting any gravitational acceleration acting on the mass, there are three forces to be taken into account:

- The force due to the moving mass, $F_m = ma = m\ddot{y}$,
- the force caused by the spring which is assumed to depend linearly on the strain applied to the spring, $F_s = sy$, and
- the force due to the velocity-linear damper, $F_d = dv = d\dot{y}$.

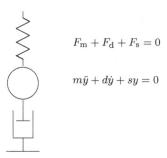

Fig. 4.18. Simple mass-spring-damper system

Since this is a closed physical system all forces add up to zero yielding the following second order differential equation, which describes the dynamic behavior of this mass-spring-damper system:

$$m\ddot{y} + d\dot{y} + sy = 0 \tag{4.8}$$

Here, m denotes the mass, d is the damper constant, and s the spring constant. Without the damping force F_d this would just be an undamped harmonic oscillator, as in the previous examples.

4.5.1 Analytical solution

This mechanical system is still simple enough to be solved analytically, which is useful as this solution can be compared later with the solutions obtained by means of an analog computer. Dividing (4.8) by m yields

$$\ddot{y} + \frac{d}{m}\dot{y} + \frac{s}{m}y = 0. \tag{4.9}$$

With the following definitions of the damping coefficient

$$\beta := \frac{d}{2m}$$

and the (undamped) angular *eigenfrequency*[51]

$$\omega_0 = \sqrt{\frac{s}{m}},$$

this can be rewritten as

$$\ddot{y} + 2\beta\dot{y} + \omega_0^2 y = 0. \tag{4.10}$$

---

[51] The eigenfrequency, sometimes also called *natural frequency*, is the frequency at which a system oscillates without any external forces acting on it.

4.5 Mass-spring-damper system

A classic approach to tackle such a differential equation is the exponential function. "Guessing"

$$y = ae^{\mu t}$$

yields

$$\dot{y} = \mu a e^{\mu t} \text{ and } \ddot{y} = \mu^2 a e^{\mu t}.$$

Substituting these into (4.10) yields

$$\mu^2 a e^{\mu t} + 2\beta \mu a e^{\mu t} + \omega_0^2 a e^{\mu t} = 0.$$

Dividing by $ae^{\mu t}$ results in the following quadratic equation

$$\mu^2 + 2\beta\mu + \omega_0^2 = 0,$$

which can be readily solved by applying the quadratic formula

$$\mu_{1,2} = -\frac{p}{2} \pm \sqrt{\frac{p^2}{4} - q}$$

with $p = 2\beta$ and $q = \omega_0^2$ yielding the solutions

$$\mu_{1,2} = -\beta \pm \sqrt{\beta^2 - \omega_0^2}. \tag{4.11}$$

With the definition of

$$\omega^2 = \omega_0^2 - \beta^2$$

(4.11) can be rearranged to

$$\mu_{1,2} = -\beta \pm i\omega.$$

The solution of (4.9) is thus given by the linear combination

$$\begin{aligned} y &= ae^{\mu_1 t} + be^{\mu_2 t} \\ &= ae^{-(\beta+i\omega)t} + be^{-(\beta-i\omega)t} \\ &= ae^{-\beta t}e^{i\omega t} + be^{-\beta t}e^{-i\omega t} \\ &= e^{-\beta t}\left(ae^{i\omega t} + be^{-i\omega t}\right). \end{aligned} \tag{4.12}$$

As known from complex analysis[52]

$$e^{i\omega t} = \cos(\omega t) + i\sin(\omega t) \text{ and}$$
$$e^{-i\omega t} = \cos(\omega t) - i\sin(\omega t).$$

Applying this to (4.12) yields

$$y = e^{-\beta t}\left(a\big(\cos(\omega t) + i\sin(\omega t)\big) + b\big(\cos(\omega t) - i\sin(\omega t)\big)\right)$$

---

[52] The function $\cos(\varphi) + i\sin(\varphi)$ is sometimes denoted by $\operatorname{cis}(\varphi)$ in the literature.

$$= \mathrm{e}^{-\beta t}\bigl((a+b)\cos(\omega t) + \mathrm{i}(a-b)\sin(\omega t)\bigr). \tag{4.13}$$

The $\mathrm{e}^{-\beta t}$ term is the damping term of this oscillating system while $a+b$ and $a-b$ are determined by the initial conditions. If it is assumed that the mass has been deflected by an angle α_0 at $t=0$, this will obviously be the maximum amplitude of its movement. If the mass is then just released at $t=0$ without any initial velocity given to it, the following initial conditions hold:

$$\alpha(0) = \alpha_0 \text{ and}$$
$$\dot\alpha(0) = 0.$$

From $\cos(\omega t) = 1$ and $\sin(\omega t) = 0$ for $t=0$ it follows that

$$\alpha(0) = (a+b) = \alpha_0.$$

Differentiating (4.13) with respect to t and applying the same arguments yields

$$\dot\alpha(0) = (a-b)\mathrm{i}\omega = 0$$

and thus $a-b=0$ for this case. So a mass released from a deflected position at $t=0$ with no initial velocity given to it, is described by

$$y = \mathrm{e}^{-\beta t}\alpha_0 \cos(\omega t),$$

which is exactly what would have been expected from a practical point of view. The position of the mass follows a simple harmonic function and its amplitude is damped by an exponential term with negative exponent.

The term

$$\omega = \sqrt{\omega_0^2 - \beta^2},$$

describing the angular eigenfrequency, yields

$$T = \frac{2\pi}{\omega}$$

for the period and is quite interesting as three cases have to be distinguished with respect to ω_0 and β:

$\omega_0 > \beta$: *Subcritical damping, (underdamped)* the mass oscillates.

$\omega_0 = \beta$: *Critical damping* – the systems returns to its position of rest in an exponential decay movement without any overshoot.

$\omega_0 < \beta$: In this case the system is said to be *overdamped*. It will return to its position of rest without any overshoot but more slowly than with critical damping.

It should be noted that the damped eigenfrequency ω is always lower than ω_0 depending on the amount of damping, which can be observed directly when the system is simulated on an analog computer, as shown in the following section.

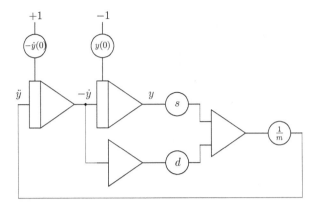

Fig. 4.19. Mass-spring-damper system

Fig. 4.20. $s=.2$, $d=.8$ **Fig. 4.21.** $s=.6$, $d=.8$ **Fig. 4.22.** $s=.8$, $d=.6$ **Fig. 4.23.** $s=.8$, $d=1$

4.5.2 Using an analog computer

Equation (4.8) is used as the basis to derive a computer setup for this mass-spring-damper system. Rearranging yields

$$\ddot{y} = -\frac{d\dot{y} + sy}{m},$$

which can be readily transformed into the computer setup shown in figure 4.19. Instead of dividing by m it is much simpler to multiply by $1/m$, although this requires a little calculation before setting the corresponding potentiometer.

Using the two potentiometers labeled $-\dot{y}(0)$ and $y(0)$ the initial conditions of the mass-spring-damper system can be set. $y(0)$ controls the initial deflection of the mass while $-\dot{y}(0)$ sets its initial velocity at the start of a simulation run. Figures 4.20 to 4.23 show the qualitative results of four simulation runs with the mass set to 1 (i.e., omitting the potentiometer following the rightmost summer altogether) and different settings for the spring and damper constants s and d. As can be clearly seen a stiffer spring requires a stronger damper in order to bring the oscillation down quickly. Furthermore, a stiffer spring results in a higher resonance frequency of the oscillating system.

The setup shown in figure 4.19 has the advantage that all three parameters can be set independently from each other. This comes at a cost, though, as two summers, one acting as an inverter, are required. Taking into account that integrators typically feature multiple inputs and that integrators and summers both

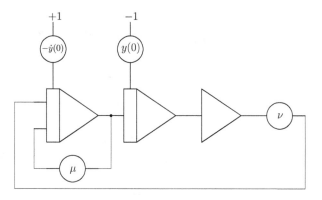

Fig. 4.24. Simplified mass-spring-damper setup

perform an implicit sign inversion, the simplified computer setup shown in figure 4.24 saves one summer.

This, however, has a drawback. The parameters s and d can no longer be set independently from m as

$$\nu = \frac{s}{m} \text{ and}$$
$$\mu = \frac{d}{m}.$$

This does match nicely with (4.9) above.

4.5.3 RLC-circuit

A physical system exhibiting similar behavior to the mass-spring-damper system can be set up from a resistor R, a capacitor C and an inductance L, as shown in figure 4.25. This system, which can also be seen as an analogue for a mass-spring-damper system, will be used here as an example for scaling an analog computer program.[53]

This circuit is described by the differential equation

$$L\ddot{Q} + R\dot{Q} + \frac{1}{C}Q = 0 \qquad (4.14)$$

with Q denoting the charge. This is structurally identical to (4.8), so everything derived so far can be applied to this RLC-circuit as well. Rearranging yields

$$\ddot{Q} = -\frac{R}{L}\dot{Q} - \frac{1}{LC}Q.$$

---

53 Cf. [Yokogawa, p. 14 et seq.].

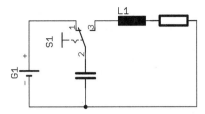

Fig. 4.25. Simple RLC-circuit

The following example is based on the computer setup shown above in figure 4.24 with $-\dot{y} = 0$. The following component values will be assumed:

$$C = 1.2 \ \mu F$$
$$L = 5.1 \ mH$$
$$R = 20 \ \Omega$$
$$E = 6.94 \ V$$

with E representing the voltage delivered by the battery in the circuit. A fully charged capacitor at $t = 0$ yields the initial condition

$$Q(0) = CE = 8.328 \cdot 10^{-6} \ As,$$

which is also the maximum amplitude of the charge variable that can be exhibited by the oscillating system.

This readily yields the following amplitude scaling factor

$$\alpha = \frac{1}{8.328 \cdot 10^{-6} \ As} \approx 120077 \ (As)^{-1}.$$

Sacrificing some of the machine unit range of $[-1, 1]$

$$\alpha = 0.1 \cdot 10^6 \ (As)^{-1}$$

will be chosen as this greatly simplifies the scaling process.

Applying this to (4.14) yields the following scaled equation:[54]

$$\widehat{\ddot{Q}} = -3.92 \cdot 10^3 \widehat{\dot{Q}} - 1.63 \cdot 10^8 \widehat{Q}$$

The initial condition $Q(0)$ must be scaled accordingly yielding (approximately)

$$\widehat{Q}(0) = 0.833.$$

---

[54] As before, the hats denote (scaled) machine variables. Units will be left out from now on to simplify notation.

With these scaled amplitudes the remaining step is time scaling. The (undamped) angular frequency[55] of the basic RLC-circuit shown in figure 4.25 is

$$\omega = \frac{1}{\sqrt{LC}} \approx 12783 \text{ s}^{-1}$$

resulting in

$$\nu \approx 2 \text{ kHz}$$

since $\nu = \omega/2\pi$. A time scaling factor of $\beta = 10^4$ is chosen which finally yields

$$\frac{\mathrm{d}^2 \widehat{Q}}{\mathrm{d}\tau^2} = -0.392 \frac{\mathrm{d}\widehat{Q}}{\mathrm{d}\tau} - 1.63\widehat{Q}.$$

The coefficients 0.392 and 1.63 can now be readily set as μ and ν in the computer setup shown in figure 4.24 with 1.63 represented by $\nu = 0.163$ feeding an integrator input weighted by 10.

This example concludes the introductory section on analog computer programming. The following chapter presents a number of useful special-purpose circuits, which also serve as additional programming examples.

---

[55] Since the undamped angular frequency is always larger than the damped frequency it is a good upper limit of what is to be expected from the circuit.

5 Special functions

As in classic digital computer programming, there are *special functions* that frequently occur in analog computer setups, such as inverse functions (like (square) roots), limits, hysteresis, etc. This chapter describes typical "library" functions which can be used as building blocks for more complex programs. As in the algorithmic domain, especially among Perl programmers, the saying "there is more than one way to do it" holds true for analog computer programs. The functions described here are typically neither the only possible implementations nor necessarily the best.

In case of doubt or when searching for other functions, the vast classic literature should be consulted. The definitive books [KORN et al. 1964] and [GILOI et al. 1963] (German) as well as [HAUSNER 1971], [GILLILAND 1967], [AMMON 1966], and [CARLSON et al. 1967] are highly recommended for further reading.

5.1 Stieltjes integral

Integration on an analog computer only works with time as the variable of integration. In some cases it is necessary to compute a more general integral of the form

$$\int_{x_0}^{x_1} f(x)\,\mathrm{d}x,$$

which can be understood as a STIELTJES integral of the form

$$\int_0^T f(t) \frac{\mathrm{d}x}{\mathrm{d}t}\, \mathrm{d}t$$

with $x = x(t)$, $x(0) = x_0$, and $x(T) = x_1$.[56] This can be easily implemented by an integrator and a multiplier as shown in figure 5.1. It should be noted that the multiplier should have a minimum zero point error since the integration amplifies any such error over time linearly, depending in the time scaling factor k_0.

5.2 Inverse functions

A very powerful technique for function generation is that of inverse functions, for example obtaining a square root based on the function $f(x) = x^2$. Recalling the basic setup of summers and integrators, the idea of a setup yielding an inverse function is simple and relies on the fact that the operational amplifier tries to keep its inverting input, SJ, at zero. Figure 5.2 shows the basic setup to generate an inverse function.

At the heart of this circuit is a function generator that forms the feedback element of an open amplifier. With an input signal of x the summing junction will be at zero if $-x = f(-y)$ with $y = f^{-1}(x)$ being the value at the output of the amplifier.

If the inverse function is not defined over the full range of $[-1, 1]$, some precautions are necessary to prevent the amplifier from overloading. It is also sometimes necessary to connect a small capacitor, typically 10s of pF, between the output of

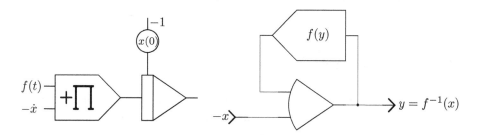

Fig. 5.1. Implementation of a STIELTJES integral

Fig. 5.2. Basic setup for generating an inverse function

---

[56] See [GILOI et al. 1963, p. 164 et seq.].

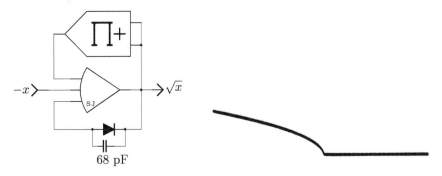

Fig. 5.3. Typical square root circuit

Fig. 5.4. Function generated by the computer setup shown in figure 5.3 with $-1 \leq x \leq 1$

the operational amplifier and its summing junction if the basic setup shows signs of instability.[57]

5.2.1 Square root

A good example of an inverse function is the square root circuit shown in figure 5.3. This is basically the same as the setup shown in figure 5.2 with two additional parts. A diode and a small capacitor are connected between the output of the open amplifier and its summing junction. The diode prevents saturation of the amplifier when the input becomes positive as in this case the input to the amplifier and the output of the multiplier will both be positive, thus driving the amplifier's output to its maximum negative level.

Although this does not harm the amplifier, driving it into saturation will generate an overflow in addition to yielding a nonsensical output value. Furthermore, the amplifier may take some time to recover from this condition. This is especially true for classic analog computers, which can exhibit recovery times of up to several seconds. Note that the diode will keep the amplifier's output a tad below zero due to its forward threshold voltage – which is about 0.7 V for a standard silicon diode and as low as 0.2 V for a SCHOTTKY diode.

The small capacitor prevents ringing when the amplifier's output is near zero. In some cases it could be omitted, depending on the behavior of the open amplifier used. Figure 5.4 shows the behavior of this circuit. The input value has been generated by a single integrator yielding a ramp running from -1 to $+1$. This ramp has also been used for the x-deflection of the oscilloscope.

---

[57] The capacitor provides a high frequency roll-off. Another approach to generating inverse functions, which can also be applied to division, will be shown in section 5.16.

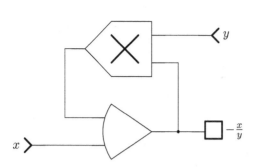

Fig. 5.5. Division as inverse function of multiplication

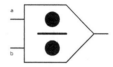

Fig. 5.6. Symbol for a divider

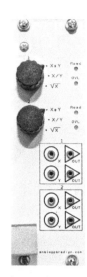

Fig. 5.7. MDS2 module containing two computing elements, each capable of performing either multiplication, division, or square rooting

5.2.2 Division

The square root circuit shown above can be easily modified to implement a division operation. Figure 5.5 shows the basic setup: With the multiplier connected to the output of the amplifier and to the input variable y, the open amplifier will drive its output towards $-x/y$, so that the output of the multiplier yields $-x$ which cancels the current delivered through the x-input and its associated input resistor to the inverting input of the amplifier.

As noted before, a small capacitor of about 68 pF should be connected between the output of the amplifier and its summing junction to prevent unstable behavior of the circuit. It should be also noted that this circuit does not perform a four-quadrant division operation.

Some systems feature pre-wired modules to compute common inverse functions such as square roots and divisions. Figure 5.7 shows an Analog Paradigm MDS2 module containing two computing elements that can be switch selected to perform either multiplication, division, or square rooting.

5.3 $f(t) = 1/t$

Some important time-dependent functions such as a time-linear ramp, $e^{-\lambda t}$ (cf. section 4.1.2), and sine/cosine (section 4.2) have already been introduced. Another useful function is $1/t$, which is a good example of how functions can be generated by solving differential equations.

The basic idea is generally to find a differential equation that has the desired function as a particular solution for certain initial conditions. The function $1/t$ is thus readily generated by the differential equation

$$\dot{x} = -x^2 \tag{5.1}$$

as the following consideration shows: Assuming $x = t^{-1}$ results in $\dot{x} = -t^{-2}$ which satisfies (5.1). The resulting computer setup is shown in figure 5.8. Since the machine time depends on the time scale factor set on the integrator it effectively computes $1/\tau$.

Figure 5.9 shows the result of a typical computer run with the initial condition $\tau^{-1}(0) = 1$, $k_0 = 100$, and OP-time set to 20 ms.

5.4 Powers and polynomials

Powers of τ and/or polynomials built from such powers are often required within a computer setup. At first glance such powers can, of course, be generated by means of multipliers fed with a linear ramp signal such as $-1 \leq \tau \leq 1$, but this is not always realistic as multipliers are expensive and analog computers, especially historic ones, typically have only few of them. Furthermore, many multipliers such as quarter square multipliers show non-negligible errors when their input values approach zero.

Consequently, it is often more advisable to generate powers of a linearly time varying variable τ by means of successive integration as shown in figure 5.10. This setup yields the powers τ, $-\tau^2$, and τ^3 for $-1 \leq \tau \leq 1$.[58]

This circuit can obviously be extended for higher powers of τ and some of the coefficient potentiometers can readily be replaced by joining inputs of the following integrator, i.e., a factor 0.2 in front of an input weighted with 10 can be replaced by feeding the value into two connected inputs, each with weight 1, etc. The successive powers of τ generated by such a setup can also be added together (with optional changes of sign by means of summers) and coefficients to form arbitrary polynomials. This is a useful technique to generate functions which

---

[58] A similar setup has already been shown in figure 2.13 in section 2.3.

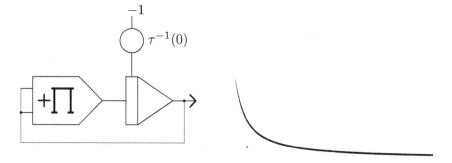

Fig. 5.8. Generating $\frac{1}{\tau}$

Fig. 5.9. Generation of τ^{-1} with $\tau^{-1}(0) = 1$ and 2.5 V per division

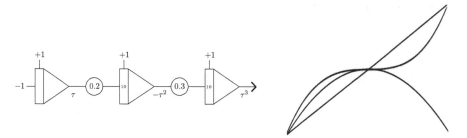

Fig. 5.10. Generating successive powers of τ

Fig. 5.11. Successive powers of τ generated by successive integrations

can be represented by a TAYLOR polynomial or the like. Figure 5.11 shows the time-dependent functions generated by the setup shown above.

5.5 Low pass filter

Sometimes, for instance when processing data gathered from experiments, a low pass filter is useful. From a rather simplified point of view such a device lets only frequencies below a certain threshold pass.[59]

Another application for such a circuit is the filtering of a (white) noise signal obtained from a noise generator. Such signals, filtered and possibly shaped into a suitable distribution (e. g., GAUSSian), are extremely useful when the behavior of mechanical systems exposed to random excitation is to be analyzed.

---

[59] A more detailed analysis would take the shape of the filter's passband as well as the induced phase-shift, etc., into account.

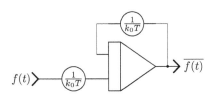

Fig. 5.12. Basic circuit of a low pass filter

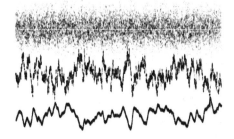

Fig. 5.13. Typical behavior of two low pass filters in series – note the mirror image relationships between successive traces due to a sign change upon integration

Although some classic analog computers such as the EAI-231(RV) or the Telefunken RA 770 feature noise generators yielding a normally-distributed output signal, such devices are rare nowadays. Noise generators intended for audio purposes feeding a low pass filter may be used instead; this is described below. If a normal distribution of the output signal is required, it can be readily obtained by using a suitable function generator.

Implementing a low pass filter on an analog computer is extremely simple as it just resembles a 1st-order time lag with the general transfer function

$$F(p) = \frac{1}{1 + pT},$$

where T represents the time constant of the delay line.[60] This setup, shown in figure 5.12, basically yields a time varying mean value of its input signal. This mean gets "better" as the time constant of the circuit increases – but this also implies that the circuit needs a longer time span to settle on the mean value.

Typically, both coefficient potentiometers are set to identical values although sometimes a certain deviation from the "ideal" setting may be advantageous depending on the desired output signal.

Figure 5.13 shows a white noise signal ranging from 0 Hz to 100 kHz obtained from a professional noise generator in the upper third of the screen. The curve in the middle is the result of applying one low pass filter of the structure shown above, while the lower trace results represents the output signal of a second low pass filter connected to the output of the first filter.

---

60 Cf. [GILOI et al. 1963, p. 308] and appendix A for more details. p is an operator implementing a time derivative.

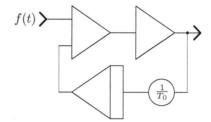

Fig. 5.14. DC block

A variation of this approach is shown in figure 5.14.[61] This implements a DC block, i.e., it removes any DC component of its input signal. $1/T_0$ determines the time constant of the filter. This circuit should be used for values $T_0 > 1$ second.

5.6 Triangle/square wave generator

Sometimes it is desirable to have a triangle or square wave signal available. Typical applications for a triangle wave are the x-deflection for an oscilloscope or plotter, while square waves are often handy to control electronic switches to display more than one curve or figure on an oscilloscope.[62]

A simple computer setup for the generation of a pair of triangle and square wave signals is shown in figure 5.15. At its heart is an integrator with its input connected to the output of a switch controlled by a comparator. This switch yields either $+1$ or -1 depending on the current slope of the triangle wave. The unlabeled coefficient potentiometer is used to set the time scale factor of the integrator.

Figure 5.16 shows a typical output signal pair generated by this circuit. With the integrator set to $k_0 = 10^3$ and the switch output connected to an input weighted with 10, a period of 400 μs is readily obtained.

5.7 Ideal diode

Diodes are very useful components as they let current pass only in one direction – at least that is what an ideal diode should do. Real diodes have some nasty traits. Most importantly they exhibit some forward threshold voltage, i.e., they won't start conducting exactly when their input crosses zero but only when it exceeds the small voltage that is an intrinsic characteristic of every diode. These threshold

---

[61] See [KORN 1966, p. 4-8].
[62] Cf. [KORN et al. 1964, ch. 9-5].

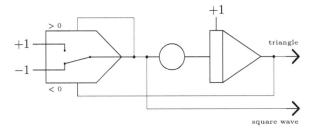

Fig. 5.15. Simple triangle/square wave generator

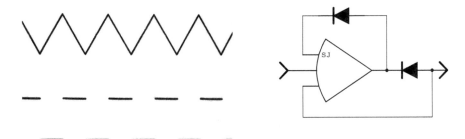

Fig. 5.16. Triangle and square wave signal generated by the circuit shown in figure 5.15

Fig. 5.17. Idealized diode circuit

voltages are typically between 0.2 V for SCHOTTKY diodes and can go up to 0.7 V for a classic silicon diode.[63]

Using an open amplifier a real diode can be readily transformed into an almost ideal diode using the setup shown in figure 5.17. The amplifier has two feedback paths. One connects its output via a diode to its summing junction, thus limiting its output to (near) zero when this diode starts conducting. The second path runs through a second diode to one of the inputs of the summer, causing it to behave like a classic inverter when this diode starts conducting. Accordingly, the output of the amplifier will be higher than the output of the overall circuit by the amount of threshold voltage of the diode used.

Figure 5.18 shows the behavior of this circuit with the diodes oriented as shown. If both diodes are reversed, the circuit will behave as shown in figure 5.19. In some cases it may be necessary to connect a small capacitor (about 68 pF) between the summer's output and its summing junction to avoid instabilities.

Obviously, a comparator with an associated (electronic) switch can also be used to implement an ideal-diode look-alike. In this case, one input of the switch

---

[63] SCHOTTKY diodes typically exhibit a larger reverse leakage current than classic silicon diodes and should be avoided in this application.

Fig. 5.18. Behavior of the circuit shown in figure 5.17

Fig. 5.19. Behavior of the circuit shown in figure 5.17 with both diodes reversed

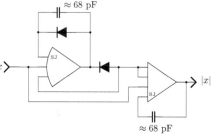

Fig. 5.20. Absolute value function with ideal diode circuit

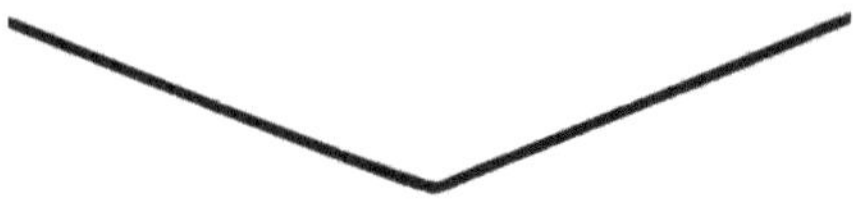

Fig. 5.21. Absolute value

is left open or is grounded while the other is connected to the input signal, which is also connected to the comparator input. Since all comparators exhibit some hysteresis to ensure stable operation, this setup may result in non-negligible errors of the diode-approximation. This setup should be avoided on a classic machine which uses electro-mechanical relays in conjunction with its comparators because relays exhibit significant variations in their closing time as well as contact bounce.

5.8 Absolute value

An idealized diode can now be used to implement an absolute value function. A typical setup is shown in figure 5.20. The left half of the circuit is the ideal diode circuit shown before. Its output feeds two paralleled inputs of a second summer. These two inputs behave like a single input weighted with 2. A third input of the second summer is directly connected to the input of the overall circuit.

When the output of the diode circuit is at zero, which is the case for a negative input value, the second summer acts as a simple inverter yielding a positive output. As soon as $x > 0$ the idealized diode starts conducting and yields $-x$. The summer on the right now computes $x - 2x = -x$ again yielding a positive output. Figure 5.21 shows the behavior of this circuit. As before, it might be necessary to add one or two small capacitors to improve the stability of the circuit.

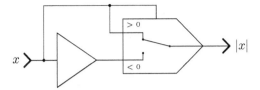

Fig. 5.22. Absolute value circuit based on a comparator

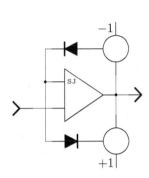 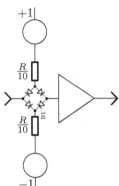

Fig. 5.23. Simple limiter circuit **Fig. 5.24.** Precision limiter circuit

If comparators with electronic switches are available an absolute value function can also be implemented in as much simple way as shown in figure 5.22. Nevertheless, due to the unavoidable tiny hysteresis of the comparator, the circuit shown before should be preferred.

5.9 Limiters

A *limiter* function limits a signal to an upper and lower bound. This is particularly useful for the simulation of mechanical systems with stops or systems exhibiting saturation effects. A very simple limiter circuit is shown in figure 5.23. At its heart is a summer which acts as an inverter as long as neither of the two forward-biased diodes conducts. This is the linear region of the limiter – its output follows the input (with the typical sign-reversal). If its output reaches the threshold of either of the two diodes shown, these will short-circuit the built-in feedback resistor of the summer and thus limits its output to the forward-bias voltage set on the two free potentiometers.

This circuit has two disadvantages. First, since it requires two free potentiometers without output buffers, it is best suited for classic analog computers which use simple voltage dividers as coefficient potentiometers. Second, the limits are

Fig. 5.25. Behavior of the precision limiter circuit

not really "flat" – they exhibit some slope that may be detrimental to the overall setup using this limiting function.

A much better limiter circuit is shown in figure 5.24. At its heart is a diode bridge controlled by two voltages representing the upper and lower limits. The two resistors should be about a tenth of the value of the input resistor of the inverter following the bridge circuit. This circuit exhibits excellent flatness of its output signal in the limited state, as shown in figure 5.25, and is highly recommended.

A very simple, yet often practical limiter circuit just employs two suitable ZENER-Diodes with their cathodes (or anodes) connected between the output of a summer or integrator and its respective summing junction.

5.10 Dead zone

Many mechanical systems, such as gear trains, linkages, etc., exhibit a phenomenon called *backlash*, which is caused by unavoidable gaps between moving parts. A *dead zone* circuit is used to model this behavior on an analog computer.

The literature contains a plethora of such circuits – the basic idea is to use a pair of suitably forward-biased diodes feeding a summer. The simplest form of such a circuit consists of just two diodes, two free potentiometers and a summer. This simplicity comes at a cost because the setup of the potentiometers is not straightforward and the diodes are far from being ideal.

The circuit shown in figure 5.26 replaces these two real diodes by idealized diodes as shown in section 5.7. The parameters r and l define the right and left break-point of the dead zone. It may be necessary to parallel the two diodes connected to the summing junctions of their associated open amplifiers with small capacitors to avoid unstable behavior of the circuit.

Figure 5.27 shows the behavior of this dead zone setup. A linear voltage ramp was used for input. The left and right gradient of the curve are determined by the weights w_r and w_l of the output summer's inputs.

Here, too, ZENER-diodes may be employed, if a very simple dead zone circuit is sufficient. In this case two of these diodes with their cathodes or anodes connected are just placed before a suitable input of a summer or integrator.

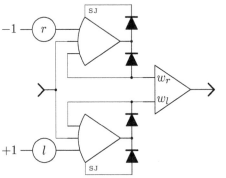

Fig. 5.26. Precision dead zone circuit

Fig. 5.27. Behavior of the dead zone circuit shown in figure 5.26

5.11 Hysteresis

A system exhibiting a *hysteresis* has some kind of "memory", i.e., its output does not only depend on its current input but also on its previous state(s). Many natural systems, such as some magnetic materials, exhibit a hysteresis effect, which is also deliberately built into devices such as simple thermostats, etc. Regarding dead zone functions, the literature contains a wealth of circuits implementing various forms of hysteresis; see for example [HOWE 1961, p. 1482 et seq.], [GILOI et al. 1963, p. 210], or [HAUSNER 1971, p. 142/145 et seq.].

The circuit shown in figure 5.29 implements rectangular hysteresis. It is simple but requires an electronic switch. The inherent hysteresis of the comparator is typically negligible but should at least be considered. The parameters a and b define the upper and lower limits of the output while α shifts the hysteresis "box" around the origin. As simple as this circuit is it has a flaw. The summer can (and will) go into overload. Accordingly, it is advisable to limit its output by a pair of ZENER-diodes connected between its output and its summing junction to suppress an overload condition which might halt the computer operation, depending on the run mode.

5.12 Maximum and minimum

Some problems require functions $max(x, y)$ or $min(x, y)$ which can be easily implemented using a comparator with electronic switch and an inverter as shown in figure 5.28.

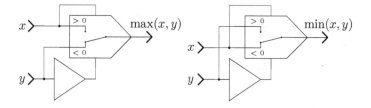

Fig. 5.28. Comparator based minimum and maximum circuits

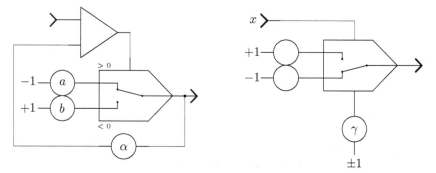

Fig. 5.29. Simple hysteresis with comparator and electronic switch

Fig. 5.30. Bang-bang circuit

5.13 Bang-bang

The *bang-bang* circuit gives one of two values at its output depending on its input voltage. This can be readily implemented by means of a comparator with an associated (electronic) switch as shown in figure 5.30. The inputs to the comparator's switch are connected to potentiometers which define the upper and lower limit of the circuit's output. One input of the comparator is connected to the input signal while the other one is fed with a threshold value γ. If this second input is grounded or omitted, the circuit will switch when the input x crosses zero. In this case the circuit effectively implements the $\mathrm{sign}(x)$ function.

Figure 5.31 shows the dynamic behavior of this simple bang-bang circuit. It is fed with a triangle wave generated by the circuit shown in section 5.6. The value γ is set to 0.75 and defines the duty cycle of the rectangular output signal, so to speak. If γ is a variable, this setup can be used as a simple modulation circuit.

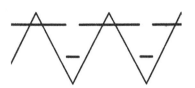 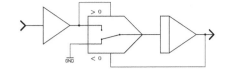

Fig. 5.31. Behavior of the bang-bang circuit

Fig. 5.32. Simple minimum holding circuit

5.14 Minimum/maximum holding circuits

During the study of stochastic systems, but not restricted to this, it is often useful to have a minimum/maximum holding circuit to store extreme values of a signal for later analysis or for use in a subsequent computer run. A plethora of circuits to accomplish this can be found in the literature, see [AMMON 1966, p. 114 et seq.] or [HAUSNER 1971, p. 146], etc. These circuits often make use of idealized diodes as described in section 5.7 requiring an open amplifier with the associated stability problems.

If the computer being used has comparators with (fast) electronic switches, very simple minimum/maximum holding circuits can be devised. Figure 5.32 shows a typical minimum holding circuit – the electronic switch connects the input of the integrator to the output of the inverting summer whenever the value at the integrators output is larger than the value at the summer's output. By interchanging the two inputs of the electronic switch, the minimum holding circuit can be transformed into a maximum holding device.

Generally, it is advisable to have the integrator set to a time scale factor as large as possible, so that it will take on new values as quickly as possible. Depending on the behavior of the integrator it may show quite significant drift with a setting like $k_0 = 10^3$ and using an input weighted by 10. Most electronic switches will also exhibit some small error increasing the drift in this setup. Connecting the free input of the switch to ground as shown can lessen this error.

Figure 5.33 shows a typical output yielded by this setup. The upper trace is a random input signal, the middle trace shows the output of the minimum holding circuit while the lower trace shows the signal controlling the electronic switch.

Nevertheless, if computations take place at high-speed, the tiny but unavoidable hysteresis of any comparator circuit will render this simple circuit pretty useless as the integrators will tend to overshoot. A much better but more complex circuit is shown in figure 5.34. At its heart is an idealized diode.[64] It is often a

---

[64] Changing the direction of both diodes will change this circuit from a peak detector to a floor detector, so to speak.

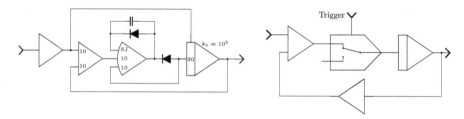

Fig. 5.33. Behavior of the simple minimum holding circuit shown in figure 5.32

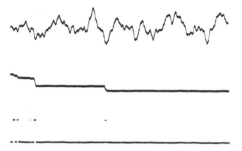

Fig. 5.34. Improved peak detector   **Fig. 5.35.** Simple sample and hold circuit

good idea to keep k_0 as low as possible with $k_0 = 10^2$ being a typical value and use inputs (or a parallel circuit of several inputs) with as large an input weight as possible. Increasing k_0 instead may yield unstable behavior.

5.15 Sample & Hold

A *sample and hold* circuit is used to sample an input signal and hold this value until another sample is requested. A simple circuit for such a device is shown in figure 5.35.[65] It relies on an (electronic) switch that is triggered by an external signal (in this case applied to its associated comparator) feeding an integrator. The input of this switch is driven by an error signal which results from the difference between the current input signal and the value stored in the integrator. When the switch is closed the integrator will drive this error signal to zero, thus yielding the last input value at its output.

It is desirable to have the integrator set to a very large time scale factor such as $k_0 = 10^3$ or larger using inputs weighted by 10 or even 100 in order to follow

---

[65] Cf. [AMMON 1966, p. 108 et seq.].

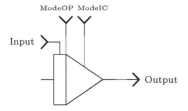

	OP	OP	IC	HALT
ModeIC	−1	+1	−1	+1
ModeOP	−1	−1	+1	+1

Fig. 5.36. Sample and hold circuit with an integrator individually controlled according to the table

the input signal with as little time lag as possible. The electronic switch may be substituted by a multiplier as long as this element has a negligible balance error.

If the analog computer being used allows individual mode control of integrators, a sample and hold circuit can be setup even more easily, as shown in figure 5.36. Many classic analog computers have a digital expansion system where the mode control lines of at least some of the computer's integrators are available. The Analog Paradigm INT4 module shown previously has four integrators, two of which can be externally controlled by means of jacks labeled ModeIC and ModeOP. The table included in the figure shows the four control signal combinations that can be applied to an integrator regardless of the current mode of operation selected on the computer's control unit.

To use an integrator as sample and hold circuit the ModeOP input should be tied to +1. Setting ModeIC to −1 will then track the input signal while applying +1 to this input will hold the last value. Typically this works well because the initial condition circuitry features an input weight of at least 100 resulting in an overall time scale factor of 10^5 with $k_0 = 10^3$.

In some cases where the tracking behavior of the integrator while being in IC-mode is undesirable, two integrators can be connected in series with the output of the first one feeding the IC input of the second. Using clever external mode controls this forms a simple *bucket brigade* circuit which yields a step-wise signal at its output.

5.16 Time derivative

In rare cases it can be necessary to compute the time derivative of a variable. However, this operation should generally be avoided at all costs because – in contrast to integration – it increases the signal noise considerably.[66] Nevertheless, when

---

[66] That is one of the main reasons why analog computers feature integrators instead of differentiators although solving differential equations is possible either way.

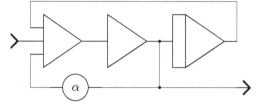

Fig. 5.37. Precision time derivative circuit

processing experimental data generating time derivatives may become necessary, e. g., to obtain acceleration from velocity.[67]

Although the integration circuit shown in figure 2.11 in section 2.3 can be converted into a differentiator by exchanging the input resistor and the feedback capacitor. Such an arrangement will exhibit excessive noise and requires at least a low pass filter to remove high frequency noise contained in the input signal. Using a summer and a free capacitor a simple yet useful differentiator can be set up by connecting the capacitor to the input of the summer. The time scale factor of this circuit is then determined by the values of the capacitor and the feedback resistor of the summer. Generally, for the input and feedback resistors R_i and R_f of the summer, the relation

$$R_i \leq \frac{R_f}{10}$$

should hold. Since the summers of an Analog Paradigm SUM8 module use 200 kΩ feedback resistors, a 4.7 nF capacitor connected to an input weighted with 10 would result in a usable circuit with a time scale factor of about 10^3.

A much more sophisticated approach is shown in figure 5.37.[68] Here the upper frequency limit can be set continuously by the parameter α according to

$$f_{\max} = \frac{k_0}{1 - \alpha}.$$

The two amplifiers in series with the coefficient potentiometer α providing feedback implement the equivalent of an open amplifier with adjustable positive feedback, in contrast to the individual negative feedbacks of each of the single amplifiers.

Figure 5.38 depicts the behavior of this circuit for different settings of α. Generally, α should be as close to 1 as possible. Values which normally work well without causing instabilities are $\alpha \approx 0.85\ldots 0.95$. Variations of this setup can be found in [HAUSNER 1971, p. 296 et seq.].

---

[67] Cf. [GILOI et al. 1963, p. 160 et seq.] for details on differentiation using analog computers.
[68] This circuit is quite similar to the DC block circuit described in section 5.5.

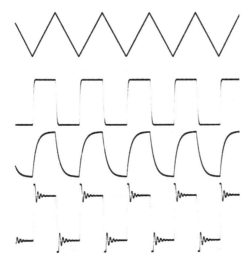

Fig. 5.38. Time derivatives of a triangle signal generated by the setup shown in figure 5.15 with upper frequency limit set correctly (second from top), too low (second from bottom), and too high (bottom)

This circuit can be used more generally to replace an open amplifier when creating an inverse function such as square rooting, division, etc. A setup for division using this approach is described in [GILOI et al. 1963, p. 168].

5.17 Time delay

Many dynamic problems exhibit delay characteristics, such as delayed neutron generation in nuclear chain reactions, transport delays in biological systems, and the transport of materials on a conveyor belt, etc. Systems like these are modelled by delay differential equations or systems thereof. Tackling such problems on an analog computer requires some means of introducing delay in a computer setup. The following sections describe three basic approaches for implementing delay functions as well as discussing some design considerations for their practical implementation.

Figure 5.39 shows the behavior of an ideal delay-circuit. A time-varying function (thick line) is delayed by a time $T > 0$ (thin line). An ideal delay with a function $f(t)$ as input yields a delayed function $f^*(t)$ as its output according to the following definition:

$$f^*(t) = \begin{cases} 0 & 0 \leq t < T \\ f(t-T) & t \geq T \end{cases} \qquad (5.2)$$

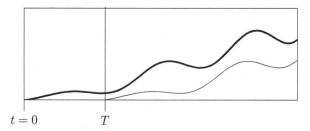

Fig. 5.39. Delaying a time-varying function

Generally, delay-circuits can be grouped into three classes depending on their behavior that results from their actual implementation:

1. Delays exhibiting discrete time and discrete values,
2. those working in discrete time and on continuous values, and finally
3. delays operating in continuous time and on continuous values.

Although delays belonging to the third class would seem to be the most desirable in an analog computer setup their implementation is quite demanding. In addition to that, these approaches exhibit a number of deficiencies (which will be discussed below) that have to be taken into account in an actual simulation setup. Delays of the first or second class are often preferred, despite their shortcomings.

5.17.1 Historic approaches to delay

In the early days of analog computation an obvious way of implementing a delay was to use a magnetic tape drive. An adjustable time delay could be implemented by either varying the tape-speed or the distance between a pair of write and read heads. Apart from the high initial cost of the tape drive itself the required frequency modulators and demodulators feeding the write heads and driven by the read heads were extremely expensive and accordingly only a few installations could afford to employ this technique.[69]

Another much simpler classical approach is based on a *capacitor wheel* which is an extension of today's sample and hold circuits. It basically consists of a large number of storage capacitors which are selected by motor-driven switches as shown in figure 5.40. Here the switches labeled 1 and 2 are driven in synchronism by a single shaft while the actual time delay is controlled by the switch labeled 3. This setup clearly belongs to the second class of delay circuits with continuous value

---

[69] More details on this can be found in [HOWE 1961, p. 225 et seq., p. 261].

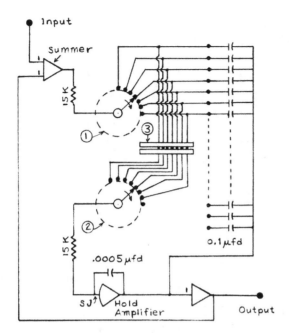

Fig. 5.40. Time delay circuit according to [THOMAS et al. 1969, p. 638]

representation but operating in discrete time. Its modern equivalent is the *bucket brigade device*, described in section 5.17.3.

The following sections will discuss the three basic classes of delay circuits and provide some practical implementation examples which are suitable for modern high-speed analog computers.

5.17.2 Digitization

Today, the apparently obvious way to implement a (variable) delay is to convert the analog signal to be delayed to a digital value using an *analog-digital-converter*, *ADC* for short. This ADC will typically contain a *sample and hold circuit* to make sure the signal does not change during conversion. The ADC output is connected to a microcontroller which stores the values into successive memory locations of a suitable *random access memory* (*RAM*). These values are then read out with suitable address offset and fed to a *digital-analog-converter*, a *DAC*, thereby achieving the desired delay, which is determined by the sampling rate and the address offset. Using off-the-shelf hardware like an Arduino®, such a digital delay can be implemented easily and cheaply. If more than one channel is to be delayed, special precautions must be taken to trigger all input sample and hold circuits in parallel and to employ similar stages following the DACs to avoid data skew.

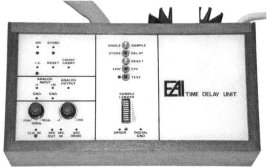

Cycle	Group I		Group II	
	IC	HALT	IC	HALT
0	✓			
1		✓		
2			✓	
3				✓
4	✓			
5		✓		
...	...	...	...	...

Fig. 5.41. Classic EAI time delay unit

Fig. 5.42. Clock control of a sample and hold circuit

The main disadvantage of this approach is that time as well as the signal values are necessarily quantized. The time discretization is determined mainly by the speed of the ADCs/DACs, the amount of available memory, and the maximum required delay time T. The signal value discretization is determined by the resolution of the ADCs/DACs. Although such circuits are readily available with 16 bits of resolution, careful layout of the printed circuit board is required. If these discretizations are acceptable, an Arduino®-based digital delay circuit can be considered.

A classic example of such a digital delay unit dating back to the late 1970s is shown in figure 5.41. If such a device is readily available, it should be given precedence over all other implementation variants because these typically require a lot of computing elements and will also exhibit signal distortion that in some cases may not be acceptable.

5.17.3 Sample and hold circuits

The sample and hold circuits shown in figure 5.43 are representative of the second class of delay systems. Such delay networks consist of an even number of individually controlled integrators (denoted by the small black stripe) grouped in two groups, labeled I and II. These two integrator groups are then alternately toggled between IC and HALT mode by means of a two-phase clock signal as shown in figure 5.42.

Each group alternately cycles through the IC- and HALT-modes, thus moving the value at the input of this integrator chain in a step-wise fashion from left to right with the delay time T depending on the clock frequency and the number of double integrator stages. Circuits like these have been used extensively to imple-

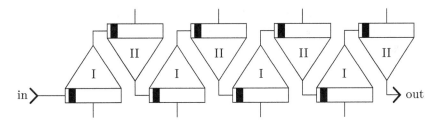

Fig. 5.43. Basic sample and hold circuit chain

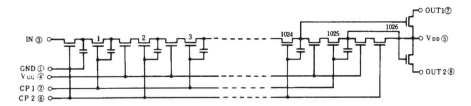

Fig. 5.44. Basic bucket brigade circuit as implemented in the MN 3007 integrated circuit (cf. [Panasonic 3007, p. 35])

ment sampled data systems and difference equations,[70] but they are not directly suitable to implement a highly granular delay circuit due to the staggering number of integrators that would be required.

A more suitable implementation of such an integrator chain is called a *bucket brigade device*, *BBD* for short. This name is quite descriptive as they are implemented by a number of capacitors connected as shown in figure 5.44.[71] These capacitors are interconnected by *field effect transistors* (*FETs*), which are controlled by a two-phase clock signal, CP1 and CP2, controlling the odd and even numbered FETs respectively. A voltage applied to the input IN is stored in the first capacitor of the bucket brigade by a clock pulse on CP2. This pulse is followed by a pulse on CP1 which transfers this value from the first capacitor to the second and so on. Since the number of capacitor stages is fixed[72] the desired delay is achieved by adjusting the frequency of the two-phase clock signal.[73]

The complementary output stage shown in figure 5.44 is noteworthy. Although one of the two output connections would suffice, the output signal is typically derived from this device by adding both output currents from OUT1 and OUT2

---

[70] Cf. [KORN et al. 1972, p. 34 et seq.].
[71] The integrated circuit MN3007 is taken as a representative example of a bucket brigade device in this context.
[72] Typical implementations contain 512, 1024, or 2048 storage capacitors.
[73] A typical device for generating this two-phase clock is the MN3101, see [Panasonic 3101].

before feeding the combined signal to an output filter to suppress any spurious high-frequency components introduced by the two-phase switching processes in the bucket brigade.

The delay obtained by a certain clock frequency is given by

$$t_{\text{delay}} = \frac{N}{2 f_{\text{clock}}},$$

where N denotes the number of bucket brigade stages and is typically of the form $N = 2^n$ with $n \in \mathbb{N}$, $8 \le n \le 11$. The clock frequency is limited by $f_{\min} \le f_{\text{clock}} \le f_{\max}$. In the case of the MN3007 integrated circuit $f_{\min} = 10$ kHz and $f_{\max} = 100$ kHz with $N = 1024$. The upper clock frequency limit is mainly due to the high capacitance of the clock inputs CP1 and CP2 whilst the lower limit is due to the inevitable leakage currents of the FET-controlled capacitors.[74]

With these constraints a MN3007 based delay circuit is nominally capable of delay times 5.12 ms $\le t_{\text{delay}} \le 51.2$ ms.[75]

5.17.4 Analog delay networks

To implement a delay circuit operating in continuous time with continuous values, first the LAPLACE transform of (5.2) is required, which can be derived as follows:[76]

$$F^*(s) = \mathcal{L}(f^*) = \int_0^\infty e^{-st} f^*(t)\, dt = \int_T^\infty e^{-st} f(t-T)\, dt$$

$$= \int_0^\infty e^{-s(t+T)} f(t)\, dt = \int_0^\infty e^{-sT} e^{-st} f(t)\, dt = e^{-sT} F(s)$$

The resulting term e^{-sT} is called the LAPLACE *delay operator*. Practical delay circuits can now be implemented by interpreting this operator as a transfer system[77] which is described by its *transfer function*. Generally, given a time-dependent input/output signal pair $y_i(t)$ and $y_o(t)$ the behavior of the transfer system can be described by

$$F(s) = \frac{y_o(s)}{y_i(s)} \quad (5.3)$$

with

$$y_o(s) = \mathcal{L}(y_o(t)) \text{ and } y_i(s) = \mathcal{L}(y_i(t))$$

---

[74] If some degree of signal degradation is acceptable, $f_{\min} = 1$ kHz may be acceptable, too.
[75] With $f_{\min} = 1$ kHz the maximum delay time can be as long as 512 ms.
[76] Cf. [CARSLAW et al. 1941, p. 7].
[77] The following sections are largely based on [GILOI et al. 1963, p. 288 et seq.].

being the *Laplace transforms* of the time varying input and output signals. Such transfer systems can be grouped into two classes:

1. Systems which can be described by ordinary differential equations and
2. systems which must be described by partial differential equations.

Only systems of the first class will be considered here. These can be described by a rational function of the form

$$F(s) = \frac{\sum_{k=0}^{m} b_k s^k}{\sum_{k=0}^{n} a_k s^k}. \tag{5.4}$$

Combining (5.3) and (5.4) and rearranging yields

$$\sum_{k=0}^{n} a_k s^k y_o(s) = \sum_{k=0}^{m} b_k s^k y_i(s).$$

Proceeding in a rather informal way, s can be interpreted as the differential operator d/dt. Replacing the Laplace transforms $y_i(s)$ and $y_o(s)$ by their corresponding time based functions then yields

$$\sum_{k=0}^{n} a_k \frac{d^k}{dt^k} y_o(t) = \sum_{k=0}^{m} b_k \frac{d^k}{dt^k} y_i(t),$$

which can now be used to derive an analog computer program for a given transfer system like the delay function.

Unfortunately, the naïve approach

$$e^{-sT} = \sum_{k=0}^{\infty} \frac{(-1)^k (sT)^k}{k!}$$

is not particularly well suited for an analog computer program since it requires an exorbitant amount of circuitry, even for a rough approximation. Ameling[78] gives some interesting additional series expansions which, unfortunately, are also not very useful for a practical implementation.

An interesting approach is to abandon the direct implementation of a power series representation of the delay operator and to use its macroscopic behavior instead to derive a suitable approximation. With $s = i\omega$ the Laplace delay operator satisfies

$$|e^{-i\omega T}| = 1,$$

---

[78] See [Ameling 1963, p. 242 et seq.].

i.e., it exhibits a constant unit gain. Furthermore, it is

$$\arg\left(e^{-i\omega T}\right) = -\omega T = -2\pi f T$$

with $\omega = 2\pi f$ as usual, so e^{-sT} yields a frequency-proportional phase shift for a pure sinusoidal input. Accordingly, this operator can be seen as an all-pass filter. Based on this observation a different approach to its implementation on an analog computer is possible.

The required unit gain forces the enumerator of (5.3) to be the complex conjugate of its denominator yielding a transfer function of the general form

$$F(s) = \frac{\sum_{k=0}^{\infty}(-1)^k a_k s^k T^k}{\sum_{k=0}^{\infty} a_k s^k T^k}.$$

The LAPLACE delay operator may now be approximated quite simply by the following first-order PADÉ approximation:[79]

$$P_1(s) = \frac{y_o(s)}{y_i(s)} = \frac{1 - \frac{sT}{2}}{1 + \frac{sT}{2}} = \frac{2 - sT}{2 + sT}$$

Cross multiplying the two fractions yields

$$y_o(s)(2 + sT) = y_i(s)(2 - sT).$$

Treating s as the differentiation operator as before this equation can be interpreted as follows:[80]

$$2y_o + T\dot{y}_o = 2y_i - T\dot{y}_i.$$

Rearranging terms yields

$$T(\dot{y}_o + \dot{y}_i) = 2(y_i - y_o). \qquad (5.5)$$

Unfortunately, this differential equation cannot readily be implemented as an analog computer program using the well-known KELVIN-feedback technique. Because there are two derivatives of degree one there is no single highest derivative to solve (5.5) for. Although time-derivatives can be generated on an analog computer,[81] this is generally not advisable. Instead, integrating over both sides yields

$$T\int_0^t (\dot{y}_o + \dot{y}_i)\,\mathrm{d}t = 2\int_0^t (y_i - y_o)\,\mathrm{d}t$$

---

[79] A thorough treatment of PADÉ approximations from a numerical point of view can be found in [SACHS 2016] and [PRESS et al. 2007, p. 245 et seq.].
[80] The argument (t) will be omitted here to avoid unnecessary clutter in the equations.
[81] See section 5.16 for a typical circuit.

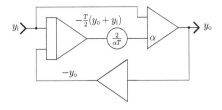
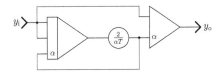

Fig. 5.45. 1st-order PADÉ approximation for the delay operator

Fig. 5.46. Simplified 1st-order PADÉ approximation

from which

$$\frac{T}{2}(y_o + y_i) = \int_0^t (y_i - y_o)\,dt$$

follows. This equation can readily be converted to an analog computer setup as shown in figure 5.45.[82] The coefficient $\frac{2}{T}$ obviously cannot be represented directly by a potentiometer setting restricted to $0 \leq \alpha \leq 1$. Accordingly, the output of the potentiometer will be connected to an input suitably weighted[83] by α of the summer on the right, as shown in the figure.

Taking into account the implicit sign inversion caused by every integrator and summer, this basic circuit can be further simplified as shown in figure 5.46. The inverting summer has been eliminated and a second path of y_i to the integrator is provided. Note that the direct feedback path from the coefficient potentiometer, too, requires an integrator input weighted with α.

The response of this 1st-degree delay operator approximation to three different input signals is shown in figure 5.47. The results are not bad given the extreme simplicity of the computer setup. In those infrequent cases where step inputs are unavoidable and less ringing of the time delay is desired, higher order PADÉ approximations can be used, as shown below for a 2nd-order system.

The 2nd-order PADÉ approximation of e^{-sT} has the form

$$P_2(s) = \frac{y_o(s)}{y_i(s)} = \frac{1 - \frac{sT}{2} + \frac{s^2T^2}{12}}{1 + \frac{sT}{2} + \frac{s^2T^2}{12}}.$$

Proceeding as above first yields

$$y_i(s)\left(1 - \frac{sT}{2} + \frac{s^2T^2}{12}\right) = y_o(s)\left(1 + \frac{sT}{2} + \frac{s^2T^2}{12}\right),$$

---

[82] A comprehensive treatment of the implementation of transfer systems on analog computers can be found in [GILOI et al. 1963, p. 288 et seq.].
[83] Typically a factor of 10, 20, or 100 will turn out to be suitable.

Fig. 5.47. Response of the 1st-order PADÉ approximation to three different input signals, $\sin(\omega t)$, $1 - e^{-\lambda t}$, and a step-function (from left to right, actual measurements)

which is then readily transformed into a differential equation by again interpreting s as the time derivative operator:

$$y_i - \frac{T}{2}\dot{y}_i + \frac{T^2}{12}\ddot{y}_i = y_o + \frac{T}{2}\dot{y}_o + \frac{T^2}{12}\ddot{y}_o.$$

Rearranging terms yields

$$\frac{T^2}{12}(\ddot{y}_o - \ddot{y}_i) = (y_i - y_o) - \frac{T}{2}(\dot{y}_o + \dot{y}_i).$$

Multiplying by $12/T^2$ yields

$$\ddot{y}_o - \ddot{y}_i = \frac{12}{T^2}(y_i - y_o) - \frac{6}{T}(\dot{y}_o + \dot{y}_i),$$

which can then be integrated twice to get rid of the time derivatives. Solving for y_o finally yields

$$y_o = y_i - \frac{6}{T}\int_0^t (\dot{y}_o + \dot{y}_i)\,\mathrm{d}t + \frac{12}{T^2}\int\!\!\int_0^t (y_i - y_o)\,\mathrm{d}t^2.$$

The resulting computer setup is shown in figure 5.48 and its behavior for three basic input signals can be seen in figure 5.49.

In cases where the behavior of these two simple PADÉ approximations (especially for the case of a step input function) are not adequate for a given problem, higher order PADÉ approximations can be found in the literature.[84] An interesting alternative approach is to drop the strict linear frequency dependent phase shift mentioned above. If some minor deviations from this frequency/phase shift relationship are allowed, good delay approximations can be derived. [GILOI et al. 1963, p. 298 et seq.] shows an interesting approach using a fourth degree approximation of the experimentally derived form

$$e^{-sT} \approx \left(\frac{19.4 - 7.63sT + s^2T^2}{19.4 + 7.63sT + s^2T^2}\right)\left(\frac{54.9 - 5.94sT + s^2T^2}{54.9 + 5.94sT + s^2T^2}\right).$$

---

[84] See [GILOI et al. 1963, p. 297], [CARLSON et al. 1967, p. 225 et seq.], or [CUNNINGHAM 1954] for a collection of suitable analog computer setups.

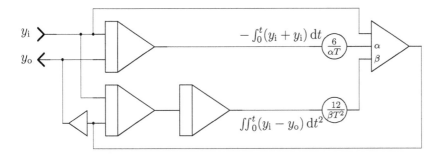

Fig. 5.48. 2nd-order PADÉ approximation for the delay operator – typical values for α and β are 20 and 100, respectively

Fig. 5.49. Response of 2nd-order PADÉ approximation to three different input signals, $\sin(\omega t)$, $1 - e^{-\lambda t}$, and a step-function (from left to right, actual measurements)

Another high-quality analog delay circuit is based on a STUBBS-SINGLE approximation which tends to get quite complicated, if high signal fidelity is to be achieved. The 4th-order STUBBS-SINGLE approximation shown in figure 5.50 is based on the transfer function

$$e^{-\tau s} = \frac{1 - \frac{1}{2}\tau s + \frac{15}{134}\tau^2 s^2 - \frac{13.55}{1072}\tau^3 s^3 + \frac{1}{1072}\tau^4 s^4}{1 + \frac{1}{2}\tau s + \frac{15}{134}\tau^2 s^2 + \frac{13.55}{1072}\tau^3 s^3 + \frac{1}{1072}\tau^4 s^4}$$

and requires four integrators, two summers, and six potentiometers.

Further details of time delay approximations can be found in various sources such as [GILOI et al. 1963, p. 294 et seq.], [STUBBS et al. 1954], [CARLSON et al. 1967, p. 225 et seq.], and [KENNEDY 1962, p. 6-5 et seq.].

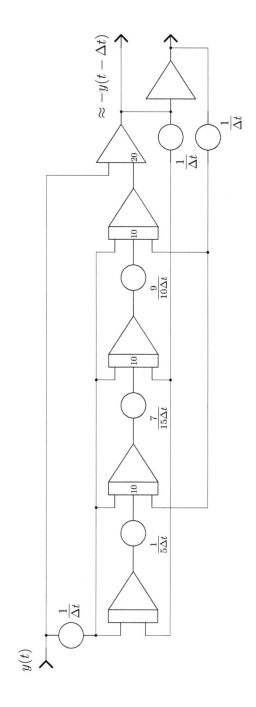

Fig. 5.50. 4th-order STUBBS-SINGLE approximation

5.18 Transfer functions

Transfer functions of the form

$$F(s) = \frac{\sum_{k=0}^{m} b_k s^k}{\sum_{k=0}^{n} a_k s^k} \tag{5.6}$$

as introduced before are an extremely useful tool in many branches of science and technology. Accordingly it is of interest to find an easy way of deriving an analog computer program from these. Stable systems satisfy $m \leq n$ and $a_k \geq 0$.

Following the reasoning outlined in section 5.17 transfer functions can be mechanized generally as shown in figure 5.51.[85]

5.19 Exponentially mapped past

It is often desirable to compute something like an arithmetic mean

$$\overline{x} = \frac{\sum_{i=1}^{n} x_i}{n}$$

for a (time) continuous variable $x(t)$. A very simple approach could look like this:

$$\overline{x} = \frac{1}{t_1 - t_0} \int_{t_0}^{t_1} x(t) \, dt$$

Although this approach works fine on an analog computer it requires fixed times t_0 and t_1, which is only practical in few cases.

To overcome this problem [OTTERMAN 1960] introduced *exponentially-mapped-past (EMP)* variables in order to extend the idea of an arithmetic mean to continuous variables in continuous time on which the classic EAI application note [EAI 1.3.2 1964] is based.[86] The basic idea is to introduce a weighting function ensuring that recent values influence the result more strongly than past values.

---

[85] Cf. [GILOI et al. 1963, p. 288 et seq.].
[86] OTTERMAN's work has its roots in [FANO 1950]. This section closely follows these sources as well as [EAI 1.3.2 1964] and [GILOI et al. 1963, p. 308 et seq.].

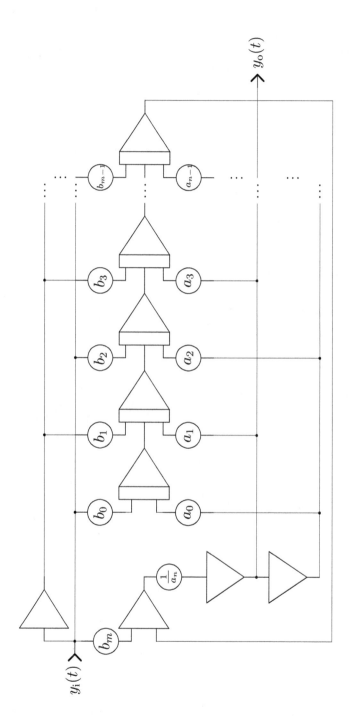

Fig. 5.51. Analog computer setup for a transfer function as shown in equation (5.6)

The following equation demonstrates this technique with the integral running from the most distant past $-\infty$ to 0 (="now"):[87]

$$\widetilde{x}(0) = \alpha \int_{-\infty}^{0} x(t) e^{\alpha t} \, dt$$

Here, α denotes a normalization factor:

$$\int_{-\infty}^{0} e^{\alpha t} \, dt = \frac{1}{\alpha}$$

Implementing this scheme on an analog computer is straightforward and guarantees that the integrator will not overload even during long operating times (given that $x(t)$ remains in suitable bounds).

This can be further generalized as

$$\widetilde{x}(T) = \alpha \int_{-\infty}^{T} x(t) e^{\alpha(t-T)} \, dt = \alpha e^{-\alpha T} \int_{-\infty}^{T} x(t) e^{\alpha t} \, dt, \qquad (5.7)$$

a convolution integral of the input function $x(t)$ and an exponentially decaying weigthing function, the derivative of which with respect to T is

$$\frac{d}{dT}\widetilde{x} = \alpha \left(-\alpha e^{-\alpha T} \int_{-\infty}^{T} x(t) e^{\alpha t} \, dt + e^{-\alpha T} e^{\alpha T} x(T) \right) = \alpha x(T) - \alpha \widetilde{x}(T). \qquad (5.8)$$

Based on (5.8) the analog computer setup shown in figure 5.52 can be directly derived. This basically implements a *leaky integrator*, which can also be seen as a low-pass RC filter. It should be noted that this only works if no exact estimation of the mean value is required during the startup time of the computation. After a step input, the output will reach 95% of the step height in the time interval $3/\alpha$. This must be taken into account for the startup time.

This approach can be extended to the calculation of an EMP variance. In the discrete case the variance is defined by

$$\sigma^2 = \frac{1}{n-1} \sum_{i=1}^{n} (x_i - \widetilde{x})^2.$$

This can be extended to continuous variables analogously to (5.7) as

---

[87] $\widetilde{x}$ denotes the EMP mean.

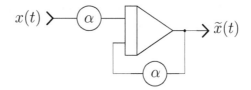

Fig. 5.52. EMP mean circuit – the parameter α determines how quickly the weigthing function "forgets" past input values

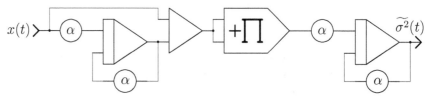

Fig. 5.53. EMP variance circuit

$$\widetilde{\sigma^2}(T) = \alpha \int_{-\infty}^{T} (x(t) - \widetilde{x}(t))^2 \, e^{\alpha(t-T)} \, dt = \alpha e^{-\alpha T} \int_{-\infty}^{T} (x(t) - \widetilde{x}(t))^2 \, e^{\alpha t} \, dt.$$

This yields the analog computer setup shown in figure 5.53.

Equally straightforward is the computation of an EMP autocorrelation $\widetilde{\rho}(\tau)$ based on

$$\widetilde{\rho}(\tau) = \alpha \int_{-\infty}^{T} x(t)x(t-\tau) e^{-\alpha(T-t)} \, dt$$

as shown in figure 5.54 where τ represents the time delay used for the correlation. The time delay function shown can be implemented using various techniques such as those shown in section 5.17.

The WIENER-KHINCHIN theorem states that the spectral decomposition of the autocorrelation function of a suitable function is given by the power spectrum of that function.[88] Thus, it is possible to compute an EMP power spectrum based on $\widetilde{\rho}(\tau)$.

The EMP FOURIER transform is defined as

$$\widetilde{F}(\omega) = \alpha \int_{-\infty}^{T} x(t) e^{-\alpha(T-t)} e^{-i\omega t} \, dt = \alpha e^{-i\omega t} \int_{-\infty}^{T} x(t) e^{-\alpha(T-t)} e^{i\omega(T-t)} \, dt.$$

---

[88] Questions regarding convergence criteria are beyond the scope of this section.

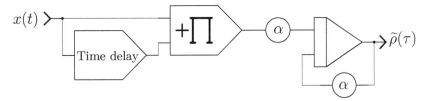

Fig. 5.54. EMP autocorrelation circuit

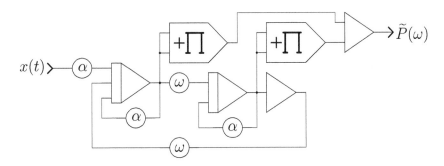

Fig. 5.55. EMP Fourier circuit

The power spectrum is $P(\omega) = |F(\omega)|^2$, i.e., in the EMP case it is

$$\tilde{P}(\omega) = \alpha^2 \left[\left(\int_{-\infty}^{T} x(t) e^{-\alpha(T-t)} \cos(\omega(T-t))\, \mathrm{d}t \right)^2 + \left(\int_{-\infty}^{T} x(t) e^{-\alpha(T-t)} \sin(\omega(T-t))\, \mathrm{d}t \right)^2 \right].$$

The corresponding analog computer setup is shown in figure 5.55. At its heart is a simple quadrature oscillator consisting of two integrators and a summer in a loop. This yields the sine and cosine components, the squares of which are summed to yield the desired output.

6

Examples

The following chapter describes a variety of problems of differing complexity which have been solved using an analog computer. Reproducing these solutions on an analog computer is not only highly instructive but also very rewarding and, last but not least, fun.

6.1 Displaying polynomials

This first example is based on the idea outlined in section 5.4 and details how a polynomial
$$p(x) = ax^3 + bx^2 + cx + d$$
with coefficients a, b, c, and d in $[-1, 1]$ can be displayed on an oscilloscope allowing a user to change the coefficients while immediately observing the effects on the polynomial.

This requires a little trick circuit to implement coefficients in $[-1, 1]$ as shown in figure 6.1, which will be used in the polynomial circuit multiple times. With $0 \leq \alpha \leq 1$ the output of this circuit will vary linearly between $[-x, x]$.

The necessary terms x, x^2, and x^3 can be obtained by successive integration over a suitable constant τ, which is linked to the OP-time of the analog computer running in repetitive mode. Figure 6.2 shows the overall program for displaying a polynomial of third degree. Ideally, an oscilloscope in x/y-mode is used with x and $p(x)$ as its respective input voltages. Figure 6.3 shows three representative screenshots obtained with this program.

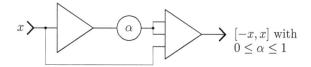

Fig. 6.1. Generating coefficients within $[-1, 1]$

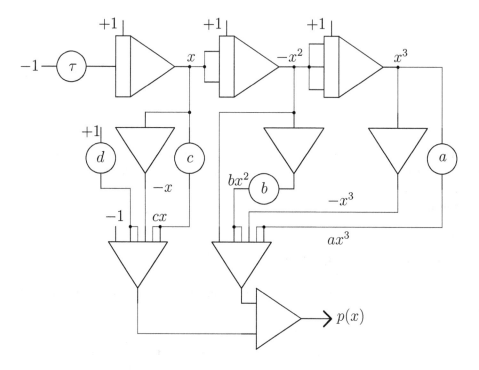

Fig. 6.2. Generating $p(x) = ax^3 + bx^2 + cx + d$ with $a, b, c, d \in [-1, 1]$

6.2 Chemical kinetics

A central problem in chemistry, biology, pharmacy, etc., is to describe and analyse the rates at which chemical reactions occur. Two general terms are important in this context. First, the *speed* of a reaction in which a substance A is transformed into a product X, a process typically denoted by $A \longrightarrow X$. The second term is the *order* of the reaction, which describes the dependency of the reaction speed on the concentration of the various substances within the reaction:

Order 0: The speed of the reaction is constant. This is typically the case when the substances involved are available in abundance.

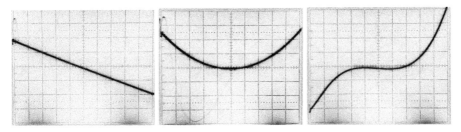

Fig. 6.3. Basic terms x, x^2, and x^3

Order 1: This describes a reaction $A \longrightarrow X$, following the equation $-\dot{a} = ka$ where k is a constant describing the speed of the reaction.

Order 2: A reaction of the form $A + B \longrightarrow X$ or $A + A \longrightarrow X$ following $-\dot{a} = kab$ or $-\dot{a} = ka^2$, if only reagent A is involved.

The simple reaction

$$A \underset{k_1}{\overset{k_2}{\rightleftarrows}} B$$

can be described by the two coupled differential equations

$$\dot{a} = -k_1 a + k_2 b \text{ and}$$
$$\dot{b} = k_1 a - k_2 b.$$

These can be readily transformed into the computer setup shown in figure 6.4,[89] which yields simulation results like those shown in figure 6.5.

A slightly more complicated reaction is described by

$$A + B \overset{k}{\longrightarrow} C,$$

yielding the equation $\dot{a} = \dot{b} = -\dot{c} = -kab$ which is easily converted into the program shown in figure 6.6. This gives the typical results shown in figure 6.7.[90]

A two step reaction system such as

$$A \overset{k_1}{\longrightarrow} B \overset{k_2}{\longrightarrow} C$$

can be described by the following set of coupled differential equations

$$\dot{a} = -k_1 a$$
$$\dot{b} = k_1 a - k_2 b \text{ and}$$
$$\dot{c} = k_2 b.$$

Figures 6.8 and 6.9 show the program and typical results for this problem.[91]

---

[89] Cf. [RÖPKE et al. 1969, p. 71 et seq.].
[90] See [RÖPKE et al. 1969, p. 51 et seq.].
[91] See [RÖPKE et al. 1969, p. 75 et seq.].

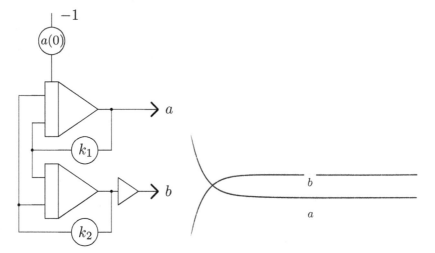

Fig. 6.4. Computer setup for the system $A \underset{k_1}{\overset{k_2}{\rightleftarrows}} B$

Fig. 6.5. Typical simulation result for the reaction $A \underset{k_1}{\overset{k_2}{\rightleftarrows}} B$

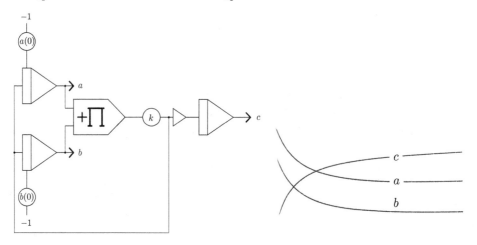

Fig. 6.6. Setup for the equation $\dot{a} = \dot{b} = -\dot{c} = -kab$

Fig. 6.7. Typical simulation result

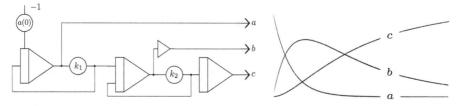

Fig. 6.8. Computer setup for the reaction $A \xrightarrow{k_1} B \xrightarrow{k_2} C$

Fig. 6.9. Simulation result for the reaction $A \xrightarrow{k_1} B \xrightarrow{k_2} C$

MICHAELIS-MENTEN *kinetics* is a classic enzyme kinetics model of the form

$$S + E \underset{k_2}{\overset{k_1}{\rightleftarrows}} \varepsilon \overset{k_3}{\longrightarrow} P + E,$$

with S, E, ε, and P denoting the concentrations of a substrate, the enzyme, the enzyme-substrate complex, and the final product of the reaction. The free enzyme on the right-hand side is available for new enzyme-substrate complexes, i.e., the enzyme is not consumed during the overall process.[92] This system is described by the following set of equations:[93]

$$\dot{s} = -k_1 es + k_2 \varepsilon$$
$$\dot{e} = -k_1 es + (k_2 + k_3)\varepsilon$$
$$\dot{\varepsilon} = k_1 es - (k_2 + k_3)\varepsilon$$
$$\dot{p} = k_3 \varepsilon$$

Setup and result are shown in figures 6.10 and 6.11.

Another complex chemical reaction model is the process of methane chlorination.[94] The basic idea is to react methane, CH_4, with chlorine, Cl_2:

$$CH_4 + Cl_2 \overset{k_1}{\longrightarrow} CH_3Cl + HCl$$
$$CH_3Cl + Cl_2 \overset{k_2}{\longrightarrow} CH_2Cl_2 + HCl$$
$$CH_2Cl_2 + Cl_2 \overset{k_3}{\longrightarrow} CHCl_3 + HCl$$
$$CHCl_3 + Cl_2 \overset{k_4}{\longrightarrow} CCl_4 + HCl$$

These reactions are governed by the reaction constants k_1, k_2, k_3, and k_4 with a, b, c, and d representing the respective concentrations of CH_3Cl, CH_2Cl_2, $CHCl_3$, and $CHCl_4$. The overall proportion of Cl_2 to CH_4 is represented by λ in the following system of coupled differential equations, which can be implemented as shown in figure 6.12:

$$\dot{a} = -k_1 a \lambda$$
$$\dot{b} = k_1 a \lambda - k_2 b \lambda = \lambda(k_1 a - k_2 b)$$
$$\dot{c} = k_2 b \lambda - k_3 c \lambda = \lambda(k_2 b - k_3 c)$$
$$\dot{d} = k_3 c \lambda - k_4 d \lambda = \lambda(k_3 c - k_4 d)$$
$$\dot{e} = k_4 d \lambda$$
$$\dot{\lambda} = -\lambda(k_1 a + k_2 b + k_3 c + k_4 d)$$

---

[92] See [RÖPKE et al. 1969, p. 105 et seq.] and [KNORRE 1971, p. 105 et seq.] for more details.
[93] In a typical scaled computer setup, k_1 is about 20.
[94] This example is based on [WAGNER 1972, p. 23 et seq.].

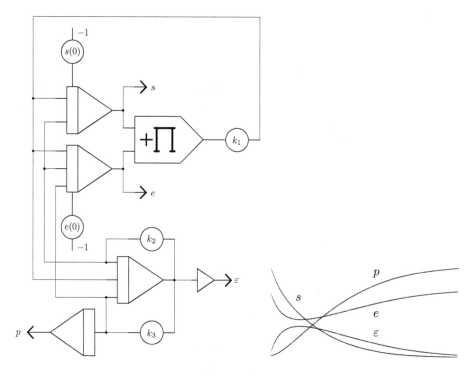

Fig. 6.10. Computer setup for the Michaelis-Menten kinetics

Fig. 6.11. Typical simulation result for the Michaelis-Menten kinetics

6.3 SEIR model

The classic *SEIR* model of an infectious disease consists of four sets of subpopulations: The set of susceptible persons S, the set of exposed persons E, the infected persons I, and the recovered (or removed, in case of death) persons R. This model[95] is described by

$$\dot{S} = -\beta SI$$
$$\dot{E} = \beta SI - \alpha E$$
$$\dot{I} = \alpha E - \gamma I$$
$$\dot{R} = \gamma I$$

---

[95] See [Schaback 2020] for a recent example of this and other systems to model COVID-19. [Bracher et al. 2021] details on the problems of predictions for epidemics.

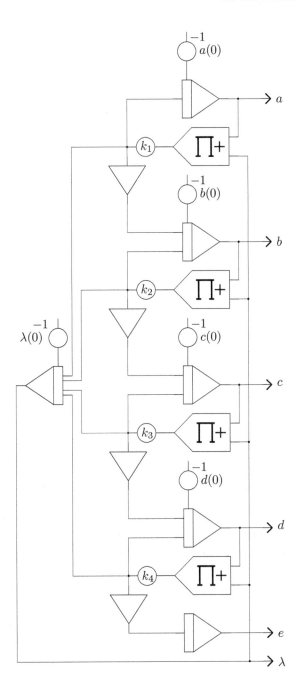

Fig. 6.12. Computer setup for the multi-stage methane chlorination reaction

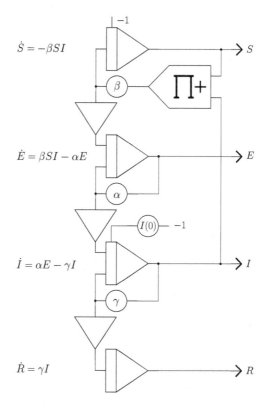

Fig. 6.13. Implementation of an SEIR model

with β representing the chance for a susceptible person to be exposed to an infected one,[96] α describing the chance of an exposed person to get ill, and γ describing the recovery (or death) rate of infected persons. The corresponding straightforward analog computer program is shown in figure 6.13.

Initially all but a few already infected persons are susceptible. This is represented by the initial conditions of the integrators yielding S and I with $I(0)$ representing the amount of persons already infected at the start of the simulation. The effects of parameter changes can be directly observed on an oscilloscope with the analog computer running in repetitive mode. A typical result is shown in figure 6.14.

---

[96] Social distancing would decrease β.

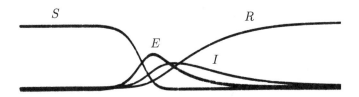

Fig. 6.14. Typical simulation result of a SEIR model

6.4 Damped pendulum with external force

The mathematical pendulum described in section 4.4 is well-behaved, because it obeys a simple linear combination of trigonometric functions once started. The following example is more complex as it not only involves an external *excitation* or *forcing function* but also includes a damping term. The forcing function $e(t)$ describes an external force to which the pendulum is subjected during its movements. A damped pendulum exposed to $e(t)$ exhibits chaotic behavior for certain sets of parameters.[97]

As before, a punctiform mass m on a massless pendulum rod of length r is assumed. The angle between this rod and a vertical line crossing its hinge is denoted by φ. As usual, g denotes the acceleration of gravity, β represents the damping factor while A is the amplitude of the forcing function. The following derivation takes a slightly different path to the one adopted in section 4.4.

With I denoting the rotational inertia it is

$$I = mr^2.$$

Since the sum of all forces in a closed system must be zero it follows that

$$I\ddot{\varphi} = \sum_i \tau_i \qquad (6.1)$$

where the τ_i denote the various torques acting in the system. These torques are

$$\tau_g = -rmg\sin(\varphi) \qquad \text{(torque due to } g\text{)} \qquad (6.2)$$
$$\tau_\beta = -\beta\dot{\varphi} \qquad \text{(torque due to } \beta\text{)}$$
$$\tau_e = A\cos(\omega t) \qquad \text{(torque due to excitation)}$$

Combining these with (6.1) results in

$$mr^2\ddot{\varphi} = -rmg\sin(\varphi) - \beta\dot{\varphi} + A\cos(\omega t)$$

---

[97] This example was inspired by the numerical simulation of such a driven pendulum shown in https://www.myphysicslab.com/pendulum/pendulum/chaotic-pendulum-en.html, retrieved on June 16th, 2019.

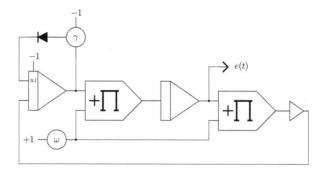

Fig. 6.15. Generating the forcing function $e(t)$

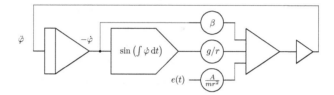

Fig. 6.16. Computer setup for pendulum simulation

which can be rearranged to give

$$\ddot{\varphi} = -\frac{g}{r}\sin(\varphi) + \frac{A\cos(\omega t) - \beta\dot{\varphi}}{mr^2}.$$

As the highest derivative is now isolated on the left-hand side, this form is ideally suited to applying KELVIN's feedback technique to derive an analog computer setup.

First, a harmonic forcing function with a variable frequency, such as the sweep circuit described in section 4.2, is required. The circuit used is shown in figure 6.15.

The sine-term in the expression (6.2) can be generated either by the subcircuit shown in figure 4.17 in section 4.4 or by a TAYLOR approximation, as already described in section 4.4.[98] The overall analog computer program for the damped pendulum subjected to a forcing function is shown in figure 6.16.

This setup invites playing with the parameters β, A/mr^2, g/r, and ω. Generally, the behavior of such an oscillatory system can be visualized best by a *phase space plot*, or *phase diagram*, which visualizes all possible states of the system. Figure 6.17 shows the chaotic behavior of the system for a certain parameter set-

---

[98] In the case of a TAYLOR approximation, φ is restricted to a certain interval, so the pendulum cannot flip over in the simulation.

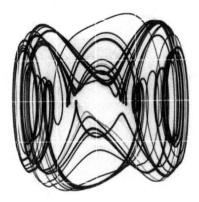

Fig. 6.17. Typical simulation result showing chaotic behavior

ting. Here $\ddot{\varphi}$ and $-\dot{\varphi}$ were plotted "against" each other, i.e., these variables were to the x- and y-inputs of the oscilloscope.

6.5 Mathieu's equation

This following section is based on another oscillating system described by the well-known MATHIEU equation – a linear homogenous second order differential equation with many applications. It can be used to describe the vibrations in an elliptic drum, the behavior of an inverted pendulum, and much more.[99]

The general form of MATHIEU's equation is

$$\ddot{y} + (a - 2q\cos(2t))\, y = 0$$

with the initial conditions $y(0) = 1$ and $\dot{y}(0) = 0$. Following [EAI 1964], the parameter a is removed by letting $a := 2q$. Defining a function

$$x(t) = 1 - \cos(2t) \qquad (6.3)$$

simplifies the equation considerably yielding

$$\ddot{y} + axy = 0. \qquad (6.4)$$

As usual, the time t as parameter of the function $x(t)$ has been omitted here and below.

---

[99] Cf. [RUBY 1996], [Yokogawa, p. 17 et seq.], and [GILOI et al. 1963, p. 196 et seq.] for more examples and details. This section is mainly based on [EAI 1964].

At first sight, one might be tempted to implement the function x using a diode function generator or a suitable approximation, but due to the fixed argument range of a function generator this approach is not viable. As is frequently the case, it is much better to generate the required function by solving a suitable differential equation. Differentiating (6.3) twice yields

$$\dot{x} = 2\sin(2t) \text{ and} \tag{6.5}$$

$$\ddot{x} = 4\cos(2t). \tag{6.6}$$

This results in the following differential equation which has (6.3) as a solution

$$\ddot{x} + 4x = 4. \tag{6.7}$$

This is demonstrably correct, as (6.5) and (6.6) can be substituted into (6.7) to give

$$4\cos(2t) + 4(1 - \cos(2t)) = 4.$$

As a first step in deriving a computer setup for (6.4) equation (6.7) first must be converted into an unscaled analog computer program. It can be written as a system of equations which quite closely resemble a program:

$$\dot{x} = \int \ddot{x}\, dt$$

$$x = \int \dot{x}\, dt \tag{6.8}$$

$$\ddot{x} = 4 - 4x \tag{6.9}$$

Obviously, $0 \leq x \leq 2$ due to (6.3), so (6.8) can be scaled by $1/2$, which must be compensated for by doubling the factor 4 in (6.9) accordingly, yielding

$$\dot{x} = \int \ddot{x}\, dt, \tag{6.10}$$

$$\widehat{x} = \int \frac{1}{2}\dot{x}\, dt, \text{ and}$$

$$\ddot{x} = 4 - 8\widehat{x}. \tag{6.11}$$

From (6.6) it is clear that $-4 \leq \ddot{x} \leq 4$, so (6.11) must be scaled by $1/4$, which will be compensated for by quadrupling $\ddot{x}$ in (6.10):

$$\dot{x} = \int 4\widehat{\ddot{x}}\, dt$$

$$\widehat{x} = \int \frac{1}{2}\dot{x}\, dt$$

$$\widehat{\ddot{x}} = 1 - 2\widehat{x} \tag{6.12}$$

Now, only $\dot{x}$ is still unscaled and must lie in the range of $[-2, 2]$ due to (6.5):

$$\widehat{\widehat{x}} = \int 2\widehat{\widehat{x}}\, dt$$

$$\widehat{x} = \int \widehat{\widehat{x}}\, dt$$

$$\widehat{\widehat{x}} = 1 - 2\widehat{x} \tag{6.13}$$

The resulting computer setup for these equations where no function exceeds the interval $[-1, 1]$ is shown in figure 6.18. Keep in mind that the output signal has been scaled down from the interval $[0, 2]$ to $[0 : 1]$ – a fact that has to be taken into account during the scaling of (6.4).

Now, equation (6.4) must be scaled accordingly. With $0 \leq x \leq 2$ a static variant $\ddot{y} + 2ay = 0$ suitable for the further scaling process is obtained. This is a harmonic oscillator with a solution like

$$y = y(0)\cos(\omega t)$$

with $\omega^2 = 2a$. The first derivative of this solution is then

$$\dot{y} = -y(0)\omega \sin(\omega t).$$

Thus, $0 \leq |\dot{y}| \leq y(0)\omega$. Restricting the parameter a to $0 \leq a \leq 10$ yields

$$\omega = \sqrt{2a} \approx 5.$$

Accordingly, a scaling factor of $1/5$ is required for $\dot{y}$. Due to the fractious behaviour of the MATHIEU equation, another factor of $1/5$ is taken into account yielding an overall scaling factor of $1/25$ and thus

$$\widehat{\dot{y}} = \frac{1}{25} \int \ddot{y}\, dt$$

$$\widehat{y} = \frac{1/5}{1/25} \int \widehat{\dot{y}}\, dt = 5 \int \widehat{\dot{y}}\, dt$$

$$\ddot{y} = 10a\widehat{y}. \tag{6.14}$$

The factor 10 in (6.14) can now be distributed over the equation. Introducing an additional scaling factor $1/10$ for a to simplify setting this parameter finally yields the program shown in figure 6.19, which allows for values $0 \leq a \leq 10$ except for the regions where the system is unstable.

Figures 6.20 shows a collection of typical solutions for MATHIEU's equation for increasing values for a. These solutions were obtained with the integrator time scale factor set to $k_0 = 10^3$ and the computer running in repetitive operation. The oscilloscope was set to 2 ms per division horizontally and 2 V per division vertically.

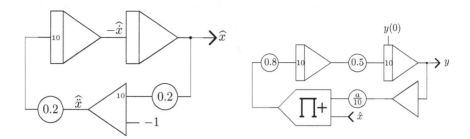

Fig. 6.18. Scaled setup for equation (6.7) **Fig. 6.19.** Scaled setup for equation (6.4)

6.6 Van der Pol's equation

The VAN DER POL equation, named after BALTHASAR VAN DER POL in 1920 as a result of his pioneering work on vacuum tubes, describes an oscillator with damping behavior which depends on the amplitude of its output. For small amplitudes the damping is negative, so the amplitude will rise until it reaches a certain threshold at which the damping term becomes positive resulting in an automatic amplitude stabilization.[100]

The basic form of the VAN DER POL equation is

$$\ddot{y} + \mu \left(y^2 - 1\right) \dot{y} + y = 0$$

which can be rearranged to give

$$\ddot{y} = -y - \mu \left(y^2 - 1\right) \dot{y}.$$

Damping is controlled by the term $y^2 - 1$, which is negative for $y < 1$, thus increasing the amplitude. If the amplitude $y > 1$, the damping term becomes positive and decreases the amplitude, thereby effectively stabilizing it.

This can be readily transformed into the unscaled computer setup shown in figure 6.21. The parameter μ controls the overall behavior of the oscillator and is assumed to be in the interval $[0, 2]$ without loss of generality.

Scaling this program is not trivial, because the parameter μ complicates things significantly. A bit of experimentation and guesstimating values shows that $-2 \leq y \leq 2$ is a good starting point. To scale this down to the interval $[-1, 1]$, the integrand of the second integrator is scaled by a factor of $1/2$. To compensate for this the output of the first multiplier must be scaled by 4, while the upper input to the summer feeding the leftmost integrator must be scaled by 2.

---

100 See [VAN DER POL et al. 1928] for more details on this type of oscillator. In this paper, VAN DER POL and his co-author J. VAN DER MARK develop an electronic model for a heart.

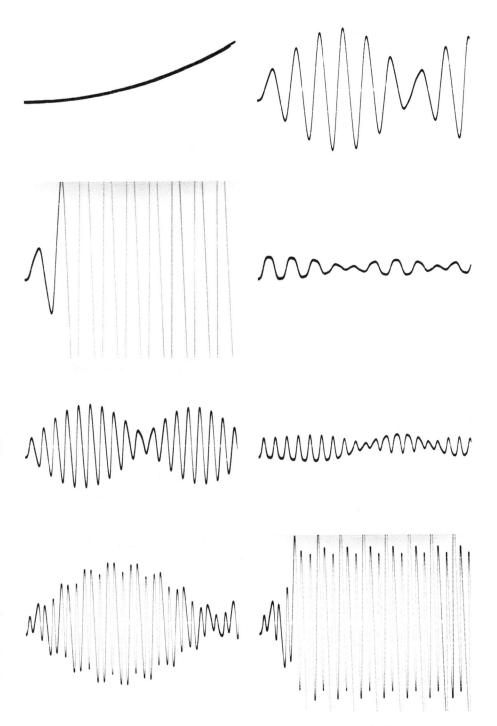

Fig. 6.20. Typical solutions of Mathieu's equation for some values $0 \leq a/10 \leq 1$

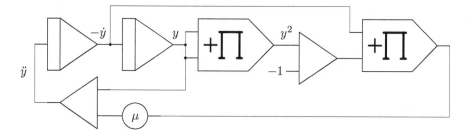

Fig. 6.21. Unscaled computer setup for the VAN DER POL equation

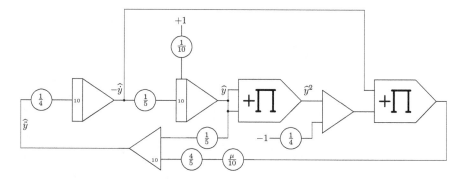

Fig. 6.22. Final computer setup for the VAN DER POL equation

The next step involves scaling of y^2, which lies in the interval $[0, 4]$. Accordingly, the scale factor of 4 at the output of the multiplier just introduced above will be removed and this will be compensated for at the input of the following summer and so forth.

Executing this process for all variables within the program yields the following set of equations with auxiliary variables y_1, y_2, and y_3, which can be readily converted into a scaled computer setup as shown in figure 6.22:[101]

$$-\widehat{\dot{y}} = -\frac{10}{4} \int \widehat{\ddot{y}} \, d\tau$$

$$\widehat{y} = -2 \int -\widehat{\dot{y}} \, d\tau$$

$$\widehat{y_1} = -\left(-\frac{1}{4} + \widehat{y}^2\right)$$

$$\widehat{y_2} = -\widehat{\dot{y}} y_1$$

---

[101] Please note that the sign-inverting feature of integrators and summers has already been taken into account in this set of equations.

Fig. 6.23. Typical phase space plot of the solution of van der Pol's differential equation

$$\widehat{\dot{y}_3} = \frac{4}{5}\widehat{y_2}\mu$$

$$\widehat{\ddot{y}} = -\left(\frac{\widehat{\dot{y}}}{5} + 10\widehat{y_3}\right)$$

As before, a phase space plot will be used to show the behavior of the oscillating system. $-\widehat{\dot{y}}$ and $\widehat{y}$ were used to create the display shown in figure 6.23. It can be seen clearly how the amplitude builds up over only one period starting from a small initial value for $\widehat{y}$ near the origin of the display. Running the computer in repetitive mode allows the observation of the influence of various settings for μ.

6.7 Generating Bessel functions

Figure 6.24 shows a classic exercise problem regarding the generation of Bessel functions on an analog computer.[102] These functions were first described by Daniel Bernoulli[103] and later generalised by Friedrich Bessel.[104] Bessel functions of the first kind are usually denoted by $J_n(t)$ and are solutions of the Bessel differential equation

$$t^2\ddot{y} + t\dot{y} + (t^2 - n^2)y = 0. \tag{6.15}$$

Sometimes these are called *cylindrical harmonics*. The parameter n in the equation above defines the *order*.[105]

---

[102] This example problem was provided by Dr. Chris Giles.
[103] 01/27/1700–03/27/1782
[104] 07/22/1784–03/17/1846
[105] See [Bowman 1958] and [McLachlan 1961] for an introduction to Bessel functions and their applications.

> ANALOG COMPUTERS
>
> Laboratory Experiment No. 5
>
> Solution of Bessel's Differential Equation
>
> By programming the Analog Computer to solve Bessel's Differential Equation, plot $J_0(x)$ and $J_1(x)$ using the Variplotter. Determine from the plot the first four zeros of $J_0(x)$ and $J_1(x)$ and compare with tabulated values of these quantities. Note that there is a relationship between $J_0(x)$ and $J_1(x)$ which should be taken advantage of in this problem.
>
> Discuss the solution and any unusual features of your solution.

Fig. 6.24. Classic exercise regarding analog computer solutions for BESSEL functions

In the following, $n = 0$ and $n = 1$ are assumed. For $n = 0$ equation (6.15) can be written as

$$\ddot{y} = -\frac{1}{t}\dot{y} - y$$

after dividing by t^2 and solving for $\ddot{y}$. This can be readily transformed into an analog computer program by applying KELVIN's feedback technique. The only thing to take into account is the term $\frac{1}{t}$ which is not well suited for an analog computer due to the pole at $t = 0$. Instead of generating this as a multiplicative term it is far more easy to directly generate the quotient $\frac{\dot{y}}{t}$ since $\dot{y} \to 0$ with $t \to 0$.

The resulting program is shown in figure 6.25. Time t has been substituted by machine time τ, which is generated using an integrator. The parameter $\dot{\tau}$ determines how fast τ rises and should be set so that it spans the whole interval from 0 to 1 during one computer run.

The relationship between $J_0(t)$ and $J_1(t)$ mentioned in the original exercise is an interesting one and can be found in [BRONSTEIN et al. 1989, p. 442] or any other standard textbook. In general

$$\frac{d}{dt}\left(t^{-n}J_n(t)\right) = -t^{-n}J_{n+1}(t)$$

holds, which implies

$$J_1(t) = -\dot{J}_0(t)$$

for the case $n = 0$. Accordingly $J_1(\tau)$ is readily available in the program as it is just $-\dot{y}$.

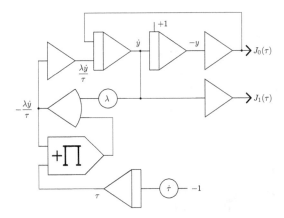

Fig. 6.25. Analog computer program for generating BESSEL functions $J_0(\tau)$ and $J_1(\tau)$

Figure 6.26 shows the overall program setup on *THE ANALOG THING*.[106] A typical result is shown in figure 6.27. Note that the program has not been properly scaled. $\dot{\tau}$ was set according to the operate-time of the machine running in repetitive mode. The central scaling factor λ was set by trial and error to get the desired result.

6.8 Solving the one-dimensional Schrödinger equation

Solving the time independent SCHRÖDINGER equation

$$\left[\frac{-\hbar}{2m}\nabla^2 + V(x)\right]\Psi(x) = E\Psi(x) \tag{6.16}$$

for a nonrelativistic particle in one dimension on an analog computer is described in this section and turns out to be pretty straightforward.[107]

The basic equation (6.16) can be rearranged into

$$-\frac{\hbar}{2m}\frac{\partial^2 \Psi(x)}{\partial x^2} + (V(x) - E)\Psi(x) = 0 \tag{6.17}$$

with

$$\hbar = \frac{h}{2\pi}$$

denoting the reduced PLANCK constant, m being the mass of the nonrelativistic particle under consideration, $V(x)$ representing the potential energy (i.e., the

---

106 See http://the-analog-thing.org.
107 This section is based on [MÜLLER 1986]. As always, ∇ represents the *nabla* or *del operator*, i.e., $\nabla = \left(\frac{\partial}{\partial x_1}, \ldots, \frac{\partial}{\partial x_n}\right)$.

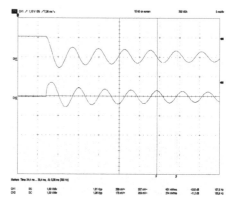

Fig. 6.26. Setup for generating $J_0(\tau)$ and $J_1(\tau)$ **Fig. 6.27.** Typical output for $J_0(\tau)$ and $J_1(\tau)$

depth of the potential well), and E being the energy of the particle. $\Psi(x)$ represents the probability amplitude, which depends on the x-coordinate of the one-dimensional system being examined. Solving (6.17) for the highest derivative yields

$$\frac{\partial^2 \Psi(x)}{\partial x^2} = \frac{2m}{\hbar}\left(V(x) - E\right)\Psi(x).$$

To solve this problem on an analog computer x will be represented by the integration time basically yielding

$$\ddot{\Psi} = \Phi\Psi \qquad (6.18)$$

with

$$\Phi := \frac{2m}{\hbar}(V - E)$$

omitting the function arguments (t) instead of (x) as usual to simplify the notation.

Equation (6.18) can be directly converted into the unscaled analog computer program shown in figure 6.28. Its input is the time-dependent function Φ describing the potential well yielding the probability amplitude Ψ as well as Ψ^2 at its outputs. The initial conditions for this function are set with the potentiometers $\dot\Psi(0)$ and $\Psi(0)$.

The computer has been run in repetitive operation with an OP-time of 20 ms and a time scale factor of $k_0 = 10^2$ set on all integrators. The input function Φ resembles a square trough and is generated with the circuit shown in figure 6.29. The integrator on the left yields a linear ramp function running from -1 to $+1$, which is fed to a series-connection of two comparators with electronic switches.

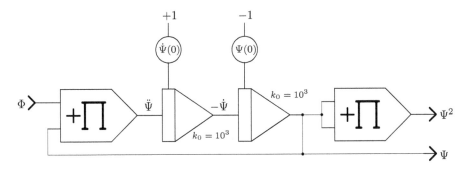

Fig. 6.28. Setup for the one-dimensional SCHRÖDINGER equation

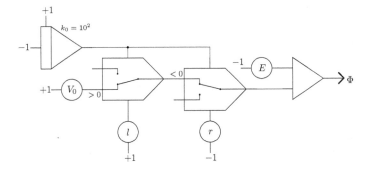

Fig. 6.29. Generating the potential well

Using the coefficient potentiometers labeled l and r the left and right position of the trough's walls can be set. The height and depth of the trough are set by the coefficients E and V_0, finally yielding Φ.

Figure 6.30 shows a typical result from an unscaled simulation run.[108] The trough parameters l and r were set to yield an approximately symmetric trough, which is shown in the upper trace. The two following curves show Ψ and Ψ^2. Here $\Psi(0)$ was assumed to be zero while $\dot\Psi(0)$ was set experimentally so that the two integrators in figure 6.28 did not go into overload.

One of the big advantages of an analog computer, namely the ease with which parameter variations can be tested, becomes very clear in this program. Varying E, V_0, $\dot\Psi(0)$, and $\Psi(0)$ gives a good sense for the behavior of the one-dimensional SCHRÖDINGER equation.[109]

---

[108] Scaling this problem is described in detail in [MÜLLER 1986].
[109] [BABERUXKI 2022] shows a more complex problem involving a GAUSSian potential.

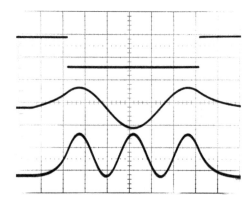

Fig. 6.30. Typical solution of the SCHRÖDINGER equation

6.9 Ballistic trajectory

The following example computes the two-dimensional ballistic trajectory of a projectile fired from a cannon taking the velocity-dependent drag, which slows down the projectile, into account.[110] This drag causes the trajectory to deviate from a simple symmetric parabola as it will be steeper on its trailing path than on its leading part. The drag $\delta(v)$ is assumed to be of the general form

$$\delta(v) = rv^2 \tag{6.19}$$

with

$$v = \sqrt{\dot{x}^2 + \dot{y}^2}.$$

This is a bit oversimplified as, according to SIACCI for example, a realistic drag function for a historic projectile has the form[111]

$$\delta(v) = 0.2002v - 48.05 + \sqrt{9.6 + (0.1648v - 47.95)^2} + \frac{0.0442v(v - 300)}{371 + \left(\frac{v}{200}\right)^{10}}.$$

Nevertheless, (6.19) will suffice for the following example. The general equations of motion of the projectile in this two-dimensional problem are

$$\ddot{x} = -\frac{\delta(v)}{m}\cos(\varphi) \text{ and} \tag{6.20}$$

$$\ddot{y} = -g - \frac{\delta(v)}{m}\sin(\varphi) \tag{6.21}$$

---

110 Cf. [KORN 1966, p. 2-7 et seq.].
111 The author would like to thank RAINER GLASCHICK for this interesting piece of history.

with g representing the acceleration of gravity, v denoting the projectile's velocity, and m being its mass. Accordingly, it is

$$\cos(\varphi) = \frac{\dot{x}}{v} \text{ and } \sin(\varphi) = \frac{\dot{y}}{v}.$$

Setting the mass $m := 1$ and rearranging (6.20) and (6.21) gives following set of differential equations:

$$\ddot{x} = -\frac{\delta(v)}{v}\dot{x}$$

$$\ddot{y} = -g - \frac{\delta(v)}{v}\dot{y}$$

The corresponding computer setup is shown in figures 6.31 and 6.32. The upper and lower halves of the circuit are symmetric except for the input for the gravitational acceleration to the lower left integrator which gives $\dot{y}$. The velocities $\dot{x}$ and $\dot{y}$ are fed to multipliers to give their respective squares, which are then summed and square rooted to get v since $\delta(v)/v = rv$ according to (6.19).

The parameters α_1 and α_2 are scaling parameters that are set to give a suitably scaled picture on the oscilloscope set to x, y-mode. Figure 6.32 shows the actual setup and parameterization yielding the result shown in figure 6.33. The initial conditions satisfy $\dot{x}(0) = \cos(\varphi_0)$ and $\dot{y}(0) = \sin(\varphi_0)$ with φ denoting the elevation of the cannon. In the screenshot shown φ_0 was set to $60°$.

6.10 Charged particle in a magnetic field

This example uses an analog computer to simulate the path of a charged particle traversing a magnetic field.[112] Basically, a particle with charge q moving in a magnetic field $\vec{B}$ is subjected to a force $\vec{F}_{\text{Lorentz}}$ which is perpendicular to both, the magnetic field $\vec{B}$ and the direction in which the particle moves, yielding

$$\vec{F}_{\text{Lorentz}} = q\left(\vec{v} \times \vec{B}\right) = q\left(\dot{\vec{r}} \times \vec{B}\right).$$

$\vec{v}$ denotes the velocity and $\vec{r}$ the position of the particle under consideration. In addition to this,

$$\vec{F}_{\text{particle}} = m\ddot{\vec{r}}$$

---

[112] This is based on [Telefunken/Particle], which was probably written by Ms. INGE BORCHARDT, who did the trajectory simulations at DESY for the high-energy particle accelerators – initially on an EAI 231RV analog computer and subsequently on a Telefunken RA 770 hybrid computer.

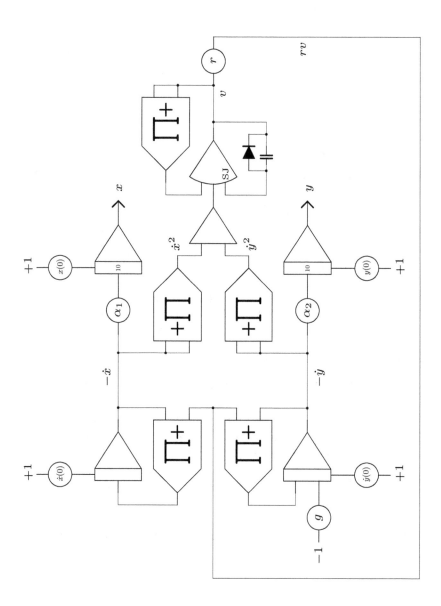

Fig. 6.31. Computer setup for the simulation of a ballistic trajectory

6.10 Charged particle in a magnetic field

Parameter	Value
$\dot{x}(0)$	0.5
$x(0)$	1
$\dot{y}(0)$	0.86
$y(0)$	1
g	0.72
r	1
α_1	0.34
α_2	0.55

Fig. 6.32. Setup and parametrization of the ballistic trajectory problem on an Analog Paradigm Model-1 analog computer

Fig. 6.33. Ballistic trajectory

holds. In addition to this, some form of friction exerting a force

$$\vec{F}_{\text{friction}} = -\mu \dot{\vec{r}}$$

is assumed to act on the particle.

Combining these three equations finally yields

$$m\ddot{\vec{r}} = \vec{F}_{\text{Lorentz}} + \vec{F}_{\text{friction}} = q\left(\dot{\vec{r}} \times \vec{B}\right) - \mu \dot{\vec{r}}. \tag{6.22}$$

It is further assumed that the particle moves in the x-y plane, which is perpendicular to the magnetic field. With e_x, e_y, and e_z denoting the unit vectors pointing into x-, y-, and z-direction, this results in

$$\vec{B} = B_z e_z \text{ and} \tag{6.23}$$
$$\dot{\vec{r}} = \dot{x} e_x + \dot{y} e_y \tag{6.24}$$

respectively. Applying a cross product to equations (6.23) and (6.24) yields

$$\dot{\vec{r}} \times \vec{B} = \dot{y} B_z e_x - \dot{x} B_z e_y.$$

Combining this with equation (6.22) finally gives

$$m\ddot{\vec{r}} = m\left(\ddot{x} e_x + \ddot{y} e_y\right) = q\left(\dot{y} B_z e_x - \dot{x} B_z e_y\right) - \mu\left(\dot{x} e_x + \dot{y} e_y\right). \tag{6.25}$$

Splitting (6.25) into components results in

$$m\ddot{x} = q\dot{y} B_z - \mu \dot{x} \text{ and}$$

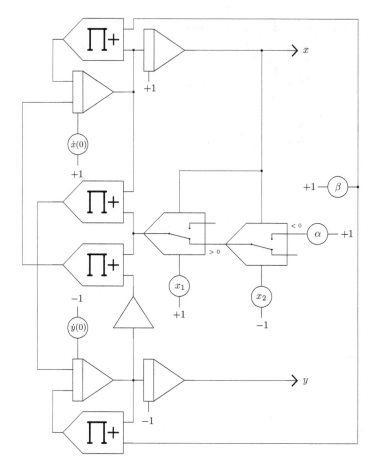

Fig. 6.34. Analog computer setup to simulate the path of a charged particle in a magnetic field

$$m\ddot{y} = -q\dot{x}B_z - \mu\dot{y}.$$

To simplify things a bit two new variables are introduced:

$$\alpha = \frac{q}{m}B_z \text{ and}$$
$$\beta = \frac{\mu}{m}$$

yielding

$$\ddot{x} = \alpha\dot{y} - \beta\dot{x} \text{ and} \quad (6.26)$$
$$\ddot{y} = -\alpha\dot{x} - \beta\dot{y}. \quad (6.27)$$

These equations can now be transformed into the analog computer program shown in figure 6.34.

Fig. 6.35. Paths of a particle not captured within the magnetic field

Fig. 6.36. Particle captured in the magnetic field

The two comparators in the middle of the figure control the area to which the magnetic field is confined. x_1 and x_2 determine the left and right coordinate enclosing the field (setting both to maximum yields a field that extends from $-1 \leq x \leq +1$). The x- and y-components of the particle's initial velocity can be controlled by the two potentiometers labeled $\dot{x}(0)$ and $\dot{y}(0)$. The outputs x and y are connected to an oscilloscope set to x, y-mode.

The results shown in figures 6.35 and 6.36 were obtained with the analog computer set to repetitive operation with an OP-time of 10 ms and time scale factors $k_0 = 10^3$ set on all four integrators.

Figure 6.35 shows two typical paths of a particle which is deflected by the magnetic field but not captured. The values for α and β were the same for both pictures. Only x_2, the coordinate at which the magnetic field ceases to act on the right side, has been increased in the second case. The particle in the left picture is fast enough to escape the narrow magnetic field while the wider field in the right picture causes the particle to be reflected.

Figure 6.36 shows the path of a particle with heavy friction. The friction is so large that the particle cannot escape the area in which the magnetic field acts.

6.11 Rutherford-scattering

In the early 20th century ERNEST RUTHERFORD performed a seminal experiment in which he bombarded a thin gold foil with α-particles.[113] He observed that the vast majority of these particles just went through the gold foil without being deflected or absorbed. Nevertheless, some of these particles were deflected through large angles with even fewer α-particles being reflected by the gold foil, a phenomenon called RUTHERFORD-*scattering*. This led to the planetary RUTHERFORD model of the atom, which then became the basis of the BOHR model.

---

[113] These are Helium nuclei and thus rather large.

The following analog computer simulation was motivated by [BORCHARDT]. It is assumed that the gold atom's nucleus is at a fixed position – specifically the point of origin of the underlying coordinate system to simplify the distance calculation for the α-particle. With the α-particle's position being (x, y) its distance to the gold nucleus is then
$$r^2 = x^2 + y^2.$$
The force exerted on the α-particle is described by COULOMB's law
$$F = k_e \frac{q_1 q_2}{r^2} \tag{6.28}$$
where $k_e = 8.9875 \cdot 10^9 \ \mathrm{Nm^2 C^{-2}}$ is COULOMB's constant and q_1, q_2 are the magnitudes of the charges involved. Since the following simulation will be a qualitative one all of these constants, including masses, can be gathered together in a new variable γ.

The force F acts proportionally to $\cos(\varphi)$ and $\sin(\varphi)$ on x and y with
$$\cos(\varphi) = \frac{x}{r} \text{ and } \sin(\varphi) = \frac{y}{r}.$$
Together with (6.28) and γ this yields the following two coupled differential equations which govern the path of the α-particle:
$$\ddot{x} = \gamma \frac{x}{r^3}$$
$$\ddot{y} = \gamma \frac{y}{r^3}$$

The setup for this problem, which is pretty straightforward, is shown in figure 6.37. It should be noted that there is an explicit additional feedback-path for the summer adding x^2 and y^2, which gives a scaling factor of $\frac{1}{2}$. Accordingly, the resulting sum satisfies $0 \leq r_* \leq 1$, so no overload condition can occur here. As mentioned above, all constants, including this factor $\frac{1}{2}$, are gathered into the parameter γ_*. A typical value for this parameter – yielding nice scattering behavior – is ≈ 0.01.

The only other free parameter is the initial height of the α-particle's trajectory $y(0)$. This can be either set manually by means of a free potentiometer connected to $+1$ and -1 or automatically as shown in figure 6.38 by means of a triangle wave generator consisting of an integrator working in conjunction with two comparators.

To run this program the analog computer is set to repetitive operation with an operation time $t_{\mathrm{OP}} = 30$ ms. The integrator of the triangle wave generator must be disconnected from the central timing control by patching its ModeIC-input to $+1$ (only the third or fourth integrator of an INT4 module can be used for that purpose), so that it will not be reset between two successive computing runs.

The picture shown in figure 6.39 was obtained with time-constants $k_0 = 100$ on the four integrators in figure 6.37 and $k_0 = 1$ in the triangle-wave generator. The camera was set to ISO 100 with an exposure time of 8 seconds.

6.11 Rutherford-scattering — 137

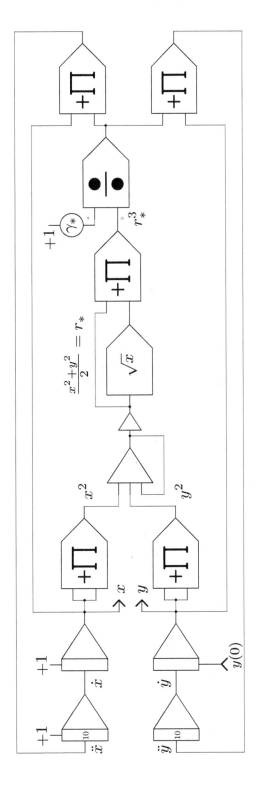

Fig. 6.37. Simulation program for Rutherford-scattering

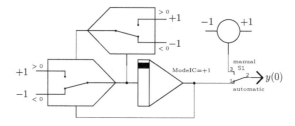

Fig. 6.38. Automatic/manual generation of $y(0)$

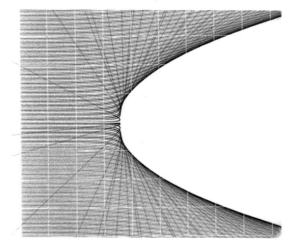

Fig. 6.39. Simulation of RUTHERFORD-scattering

6.12 Celestial mechanics

Simulating (or determining) the movement of three (celestial) bodies with given parameters, such as masses, gravitational constant, initial conditions for the bodies' coordinates and velocities, is known as the *three-body problem*.[114] The first one to tackle this problem was Sir ISAAC NEWTON in the first book, section XI, of his *Principia*, which was considered "the most valuable chapter that was ever written on physical science" according to Sir GEORGE BIDDELL AIRY.[115]

Although an analytical solution is possible in the simpler case of two bodies, JOSEPH-LOUIS LAGRANGE was able to give some particular solutions for the much more difficult three-body problem. Since these are beyond the scope of this section, the reader is referred to the standard texts [BATTIN 1999, p. 365 et seq.] and

---

[114] This is a special case of the n-body problem, which will not be investigated further here.
[115] Cf. [MOULTON 1923, p. 363].

[MOULTON 1923, p. 277 et seq.] for an analytical treatment of this problem and some background information. However, using an analog computer, it is straightforward to investigate the three-body problem.

In this section a simple universe consisting of two suns with fixed positions $\vec{r}_1 = (x_1, y_1)$ and $\vec{r}_2 = (x_2, y_2)$, masses $m_1 = m_2 = 1$, and a satellite of negligible mass criss-crossing between these two celestial bodies will be simulated. This system can be readily described by the following set of coupled differential equations:

$$m_1 \ddot{\vec{r}}_1 = Gm_1 m_2 \frac{\vec{r}_2 - \vec{r}_1}{\left| \vec{r}_2^{\,2} - \vec{r}_1^{\,2} \right|^3} + Gm_1 m_3 \frac{\vec{r}_3 - \vec{r}_1}{\left| \vec{r}_3^{\,2} - \vec{r}_1^{\,2} \right|^3}$$

$$m_2 \ddot{\vec{r}}_2 = Gm_2 m_3 \frac{\vec{r}_3 - \vec{r}_2}{\left| \vec{r}_3^{\,2} - \vec{r}_2^{\,2} \right|^3} + Gm_2 m_1 \frac{\vec{r}_1 - \vec{r}_2}{\left| \vec{r}_1^{\,2} - \vec{r}_2^{\,2} \right|^3}$$

$$m_3 \ddot{\vec{r}}_3 = Gm_3 m_1 \frac{\vec{r}_1 - \vec{r}_3}{\left| \vec{r}_1^{\,2} - \vec{r}_3^{\,2} \right|^3} + Gm_3 m_2 \frac{\vec{r}_2 - \vec{r}_3}{\left| \vec{r}_2^{\,2} - \vec{r}_3^{\,2} \right|^3} \quad (6.29)$$

The vectors $\vec{r}_i$, $1 \leq i \leq 3$ describe the positions of the different bodies with masses m_i while G is the gravitational constant.

Assuming that the two suns are stationary, only equation (6.29) has to be implemented on the analog computer. Since the suns' masses are equal to 1 the satellite's mass cancels out, yielding

$$\ddot{\vec{r}}_3 = \frac{G}{\left| \vec{r}_1^{\,2} - \vec{r}_3^{\,2} \right|^3}(\vec{r}_1 - \vec{r}_3) + \frac{G}{\left| \vec{r}_2^{\,2} - \vec{r}_3^{\,2} \right|^3}(\vec{r}_2 - \vec{r}_3). \quad (6.30)$$

Introducing two distance terms

$$\Delta r_{13}^3 = \left(\sqrt{(x_1 - x_3)^2 + y_3^2} \right)^3 \quad \text{and} \quad (6.31)$$

$$\Delta r_{23}^3 = \left(\sqrt{(x_2 - x_3)^2 + y_3^2} \right)^3 \quad (6.32)$$

and splitting (6.30) into its x- and y-components yields the following two equations which describe the satellite's trajectory in Cartesian coordinates:

$$\ddot{x}_3 = (x_1 - x_3)\frac{G}{\Delta r_{13}^3} + (x_2 - x_3)\frac{G}{\Delta r_{23}^3} \quad \text{and} \quad (6.33)$$

$$\ddot{y}_3 = -y_3 \left(\frac{G}{\Delta r_{13}^3} + \frac{G}{\Delta r_{23}^3} \right). \quad (6.34)$$

These equations can now easily be implemented on an analog computer. Figure 6.40 shows the partial computer setup yielding the distance terms (6.31) and (6.32).[116]

---

[116] The two square root units followed by a multiplier each could be replaced by two diode function generators, if these are available.

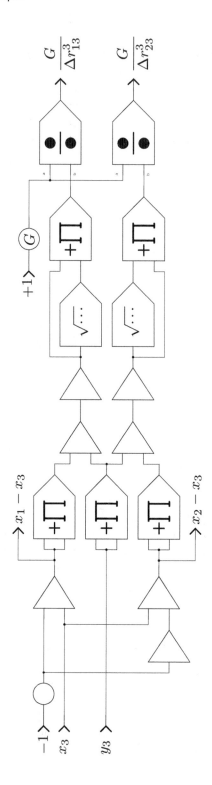

Fig. 6.40. Distance terms

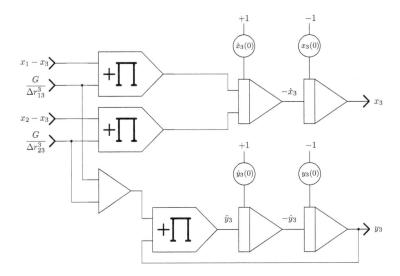

Fig. 6.41. Implementation of equations (6.33) and (6.34)

Mechanizing (6.33) and (6.34) is straightforward, too, as shown in figure 6.41. Since $y_1 = y_2 = 0$ the generation of y_3 does not require complex distance terms like the partial circuit yielding x_3, thereby simplifying the overall setup considerably.

Running this problem on an analog computer is fascinating due to the non-periodic behavior of the satellite's trajectory.[117] Figure 6.42 shows a family of trajectories of the satellite with the analog computer running in repetitive mode. The time scale factor k_0 of the integrators was set to 10^5, the operation time was 10 ms. The initial conditions were $x_1 = 0.5$, $x_2 = -0.5$, $x_3(0) = 0.1$, and $y_3(0) = 0.2$. Figure 6.43 shows a similar family of trajectories for a larger value of G.

6.13 Bouncing ball

The simulation of a ball bouncing in a box described in the following section is not only interesting in itself but also ideally suited for setting up a fascinating display for exhibitions, schools, etc. Using high-speed integration and repetitive operation of the analog computer, a flicker-free oscilloscope display of the ball's path can be easily obtained. This allows one to change the parameters of the simulation manually and directly observe the effects of these changes.[118]

---

[117] See [LIAO 2013].
[118] This section was inspired by [Telefunken/Ball].

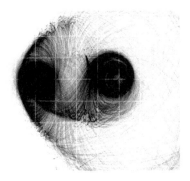 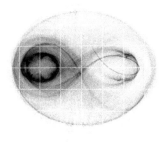

Fig. 6.42. Typical satellite trajectory for a small value of G

Fig. 6.43. Trajectory for a larger value of G

Figure 6.44 shows the basic idea of the simulation: The bounding box is assumed as the rectangle $[-1, 1] \times [-1, 1]$ and the position of the ball within this box is described by the coordinates (x, y). At the start of a simulation run the ball has two initial conditions: an initial velocity $v(0)$ (of which only the x-component will be given) and an initial y-position $y(0)$, set to 1.

The x- and y-components of the ball's position are completely independent so they can be generated by two separate sub-programs. The x-component of the ball's velocity is assumed to decrease linearly over time. The actual x-position of the ball is derived by integrating over its velocity. Every time the ball hits the left or right wall of the box, it changes its direction, i.e., it is reflected by the wall.

The y-component is that of a free-falling ball bouncing back elastically when hitting the floor. Figure 6.45 shows those two variables over time as displayed on an oscilloscope: Starting from the left, the computer is first in IC-mode. As soon as the OP-mode starts, the ball begins to drop until it hits the floor while the x-component increases (trace in the middle) with decreasing velocity until it reaches the right wall, etc.

Figure 6.46 shows the computer setup yielding the x-component of the ball's position. The leftmost integrator generates the velocity component v_x of the ball which starts at $+1$ and decreases linearly controlled by the Δv-potentiometer.[119] It should be noted that this setup differs slightly from the triangle generator shown in figure 5.15 in section 5.6 by the introduction of a second comparator with an associated electronic switch. This yields a stable comparison value of ± 1 for the first comparator since its associated switch no longer yields ± 1 at its output, because it is driven by the leftmost integrator in this setup.

---

[119] The diode at the output of the integrator is not really necessary, but it makes parameter setup easier as it prevents a negative velocity at the cost of introducing a slight voltage drop determined by its intrinsic forward threshold voltage.

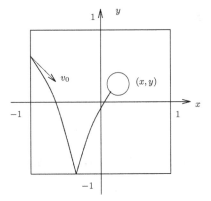

Fig. 6.44. Movement of the bouncing ball

Fig. 6.45. x- and y-component of the bouncing ball

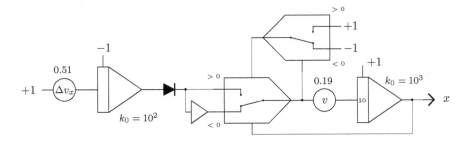

Fig. 6.46. Generating the x-component

The output of this circuit is fed to the x-input of an oscilloscope set to x, y-display mode. The time scale factors of the integrators in this example were set to 10^2 and 10^3 respectively.

The y-component of the ball in flight can be described by

$$\ddot{y} = -g + d\dot{y} \qquad (6.35)$$

where d denotes the damping coefficient due to air friction. This friction is assumed to be proportional to the ball's velocity which is a significant simplification.[120] When the ball hits the ceiling or floor of the box, an additional term is introduced which is proportional to the depth of penetration:

$$\ddot{y} = -g + d\dot{y} \begin{cases} +c(-y-1) & \text{if } y < -1 \\ -c(y-1) & \text{if } y > 1. \end{cases}$$

The case $y > 1$ – when the ball hits the ceiling of the box – cannot happen in this simulation as the vertical component of the ball's initial velocity is assumed

---

[120] Cf. section 6.9 for a more realistic model of drag.

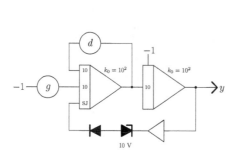

Fig. 6.47. Generating the y-component

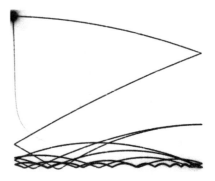

Fig. 6.48. Trace of the bouncing ball

to be zero. At the beginning of a simulation run, the ball begins a free falling trajectory controlled by (6.35). When the ball hits the floor, it will rebound due to elastic collision described by the term $c(-y - 1)$ with a constant c controlling the elasticity.

Figure 6.47 shows the corresponding computer setup. The leftmost integrator integrates over the gravity g and takes the friction d into account. It yields the negative y-component v_y of the ball's velocity. The integrator on the right integrates over v_y yielding the actual y-position.

The implementation of the elastic rebound is extremely simple and consists of two diodes. The "active" element here is the 10 V ZENER-diode which starts conducting when the ball hits the floor of the box (ignoring the small bias voltage of the diodes). The diode on the left makes sure that the ZENER-diode won't conduct while the ball is within the box. Since the rebound effect is rather violent the output of the two diodes connected in series is directly connected to the summing junction input (SJ) of the first integrator. The time scale factors of the integrators are both set to $k_0 = 10^2$.

With settings as denoted in the circuit diagrams and the computer set to repetitive operation with an OP-time of 20 ms and short IC-time, a display such as that shown in figure 6.48 can be easily obtained. Because the oscilloscope used here had no blanking-input the return of the beam to the upper left corner is faintly visible. If a blanking-input is available, it can be connected to a trigger output of the analog computer's control unit to avoid this artefact.

6.14 Zombie apocalypse

How could one not like zombie movies? It was about time that mathematicians – ROBERT SMITH? and his collaborators – shed some light on zombie attacks from a mathematical point of view.[121] The basis for a description of a zombie apocalypse are the well-known VOLTERRA-LOTKA *differential equations* modelling a closed eco-system, which were derived in the late 19th/early 20th century by ALFRED JAMES LOTKA and VITO VOLTERRA.

The mathematical model used below is derived from this classic set of differential equations:

$$\frac{dh}{dt} = \alpha h - \beta hz \qquad (6.36)$$

$$\frac{dz}{dt} = \delta hz - \gamma hz - \zeta z \qquad (6.37)$$

Here h and z represent the number of humans and zombies respectively with the initial conditions $h(0)$ and $z(0)$. The other parameters are:

α: Growth rate of the human population (birth rate).
β: Rate at which humans are killed by zombies (some kind of a *capture cross section*).
δ: Growth factor of zombie population due to zombies transforming humans into zombies.
γ: Rate at which zombies are killed by humans.
ζ: Normal "death" rate of the zombie population.

Figure 6.49 shows the resulting qualitative (i.e., unscaled) program for equations (6.36) and (6.37). A typical simulation result obtained with this setup is shown in figure 6.50. The computer was run in repetitive mode with the time scale factors of the integrators set to $k_0 = 10^3$. Additionally, all integrator inputs had a weight of 10 further speeding up the simulation by another factor of 10. The OP-time was set to 60 ms. The oscilloscope, relying on its built-in time-deflection, was explicitly triggered by one of the trigger outputs of the CU to obtain a stable display.

The parameters used were derived experimentally by manually changing the coefficients until an oscillatory behavior was obtained. The output shown corresponds to $h(0) = z(0) = 0.6$, $\alpha = 0.365$, $\beta = 0.95$, $\delta = 0.84$ (very successful zombies, indeed), $\gamma = 0.44$, and $\zeta = 0.09$. As in most predator-prey-systems it is quite difficult to find a parameter setting which gives oscillatory behavior with stable minima and maxima amplitudes. Since neither species becomes extinct with this particular parameter set there is plenty of scope for many Zombie movies in years to come.

---

121 See [SMITH? 2014] and [MUNZ 2014].

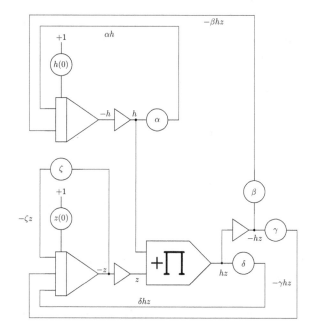

Fig. 6.49. Analog computer program for equations (6.36) and (6.37)

Fig. 6.50. Results of a typical zombie simulation

6.15 Rössler attractor

An electronic analog computer is ideally suited to studying continuous-time dynamic systems of which those exhibiting chaotic behavior are not only fascinating but also yield aesthetically pleasing phase space plots. The following sections describe a number of such systems; the first of these was developed and studied in 1976 by OTTO RÖSSLER.[122] This system is described by the three coupled differential equations

$$\dot{x} = -(y + z),$$
$$\dot{y} = x + ay, \text{ and}$$

---

[122] See [RÖSSLER 1976].

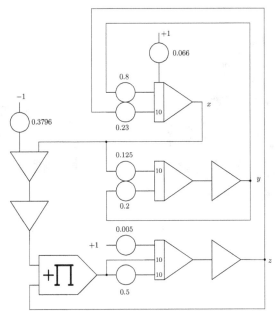

Fig. 6.51. Setup for the RÖSSLER attractor

$$\dot{z} = b + z(x - c)$$

where $a = 1/5, b = 1/5$, and $c = 5.7$. A remarkable property of this particular system is that only one of these equations is non-linear.

Since the variables in these equations will exceed the valid range $[-1, 1]$ of the machine units, the equations must be scaled before setting up the computer. Due to the non-linear nature of the problem, this scaling process is rather cumbersome and is best done in a multi-step fashion. The final set of scaled coupled DEQs for this problem looks like this:

$$\dot{x} = -0.8y - 2.3z$$
$$\dot{y} = 1.25x + a^*y$$
$$\dot{z} = b^* + 15z(x - c^*)$$

with $a^* = 0.2, b^* = 0.005$, and $c^* = 0.3796$. The resulting computer setup is shown in figure 6.51 while figure 6.52 shows an x-y-plot of the attractor, photographed with ISO 100 and a time scale factor $k_0 = 10^3$ set on all three integrators.

A particularly beautiful picture can be obtained with a simple (static) 3d-projection of the attractor. One input of the oscilloscope, which is set to x, y-mode, is directly fed by x while the other input is fed from a summer yielding

$$y^* = -(y\sin(\varphi) + z\cos(\varphi)),$$

where the sine/cosine terms are directly set by coefficient potentiometers as shown in figure 6.54. Using these two coefficient potentiometers the angle of view is freely

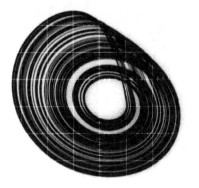 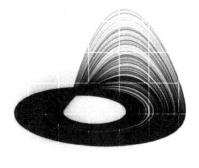

Fig. 6.52. x-y-plot of the RÖSSLER attractor   **Fig. 6.53.** Projection of a RÖSSLER attractor

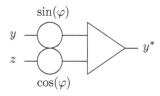

Fig. 6.54. Projection circuit

adjustable. Figure 6.53 shows a typical projection of the attractor obtained with this setup.

6.16 Lorenz attractor

Another, even more famous, chaotic system is the intriguing LORENZ attractor developed in 1963 by EDWARD NORTON LORENZ as a simplified model for atmospheric convection and first described in [LORENZ 1963].[123] Although this seminal work was performed on a digital computer, a Royal McBee LGP-30, the LORENZ attractor is, of course, ideally suited to being implemented on an analog computer as will be shown below.

This dynamic chaotic attractor is described by the following system of three coupled differential equations:

$$\dot{x} = \sigma(y - x)$$
$$\dot{y} = x(\rho - z) - y$$

---

[123] More information about the construction of chaotic attractors in general as well as this attractor in particular may be found in [KUEHN 2015, p. 468 et seq.].

$$\dot{z} = xy - \beta z$$

The parameters are $\sigma = 10$, $\beta = \frac{8}{3}$, and $\rho = 28$. Obviously, this set of DEQs must be scaled in order to be implemented on an analog computer. The resulting scaled equations look like this:

$$x = \int (1.8y - x)\,\mathrm{d}t + C$$

$$s = 1 - 2.678z$$

$$y = \int (1.5556xs - 0.1y)\,\mathrm{d}t$$

$$z = \int (1.5xy - 0.2667z)\,\mathrm{d}t.$$

C denotes the initial condition of the integrator yielding x and is not critical. Taking into account that every summer and integrator of an analog computer performs an implicit change of sign and further noting that $xy = -x(-y)$, these equations can be further simplified saving two inverters in the resulting computer setup:

$$-x = -\int (1.8y - x)\,\mathrm{d}t + C$$

$$-z = -\int (1.5xy - 0.2667z)\,\mathrm{d}t$$

$$s = -(1 - 2.68z)$$

$$r = -xs$$

$$-y = -\int (1.536r - 0.1y)\,\mathrm{d}t.$$

The corresponding schematic is shown in figure 6.55.[124]

Depending on the time scale factor k_0 set for the integrators either an x,y-plotter or an oscilloscope may be used to give a graphical representation of this chaotic attractor. Using the simple circuit for obtaining a static 3d-projection as shown in the preceding section in figure 6.54, the displays shown in figures 6.56 and 6.57 were obtained.

6.17 Another Lorenz attractor

In 1984 EDWARD NORTON LORENZ described another chaotic attractor that did not become as famous as the one described above but is nevertheless interesting to

---

[124] Although not fully consistent with the scaling of the equations, a value of 0.125 instead of 0.2667 at the feedback potentiometer of the second integrator has shown to yield great results.

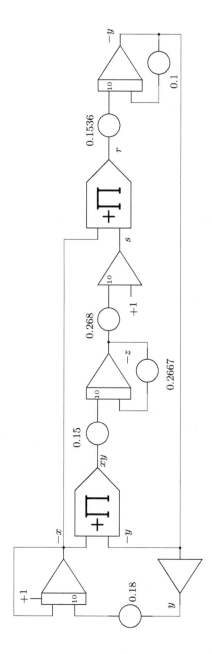

Fig. 6.55. Computer setup for the LORENZ attractor (the value of 0.2667 in the feedback of the second integrator can be decreased to about 0.125 to yield a nice result)

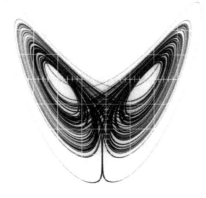

Fig. 6.56. Classic display of the Lorenz attractor

Fig. 6.57. Different angle of view on the Lorenz attractor

implement on an analog computer. It is described by the three coupled differential equations

$$\dot{x} = -y^2 - z^2 - ax + af,$$
$$\dot{y} = xy - bxz - y + g, \text{ and}$$
$$\dot{z} = bxy + xz - z$$

with the parameters $a = \frac{1}{4}$, $b = 4$, $f = 8$, and $g = 1$.[125]

Scaling this system is pretty straight-forward as a quick numerical experiment shows that $|x|$ and $|y|$, $|z|$ are bounded by 2. The resulting analog computer program is shown in figure 6.58. Tweaking the parameters is quite interesting and a typical set of solutions is shown in figure 6.59.

6.18 Chua attractor

A much more complex chaotic system, the *Chua oscillator*, is described in this section. This system was discovered in 1983 by Leon Ong Chua and is a classic example of an electronic circuit exhibiting chaotic behavior. This oscillator generates a unique and particularly beautiful attractor called the *Double Scroll attractor*. At its heart is a (hypothetical) nonlinear device called a *Chua diode*.

The mathematical description of this particular oscillator is based on three coupled differential equations of the form

$$\dot{x} = c_1(y - x - f(x)), \tag{6.38}$$

---

[125] See [Lorenz 1984].

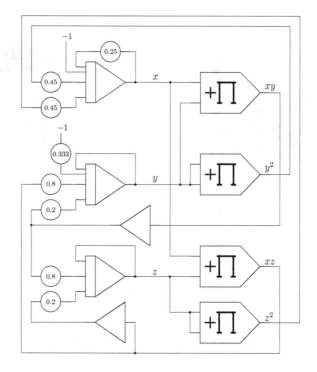

Fig. 6.58. Program for the chaotic LORENZ-84 system

$$\dot{y} = c_2(x - y + z), \text{ and} \qquad (6.39)$$
$$\dot{z} = -c_3 y \qquad (6.40)$$

where $f(x)$ describes the behavior of the CHUA diode and is defined as

$$f(x) = m_1 x + \frac{m_0 - m_1}{2}(|x+1| - |x-1|). \qquad (6.41)$$

The standard values for the parameters are

$$c_1 = 15.6,$$
$$c_2 = 1,$$
$$c_3 = 28,$$
$$m_0 = -1.143, \text{ and}$$
$$m_1 = -0.714.$$

Scaling this system of coupled differential equations is not simple due to its pronounced non-linear behavior. The ranges of the various terms were determined by computing example solutions on a digital computer – a technique that was already used in the 1960s. These results were then used to guide the scaling process.

6.18 CHUA attractor — 153

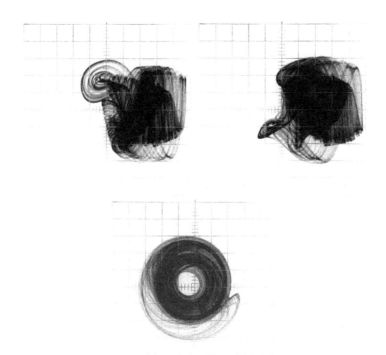

Fig. 6.59. Typical behavior of the LORENZ-84 system, the three screenshots show x vs. y, x vs. z, and y vs. z

To simplify the overall process, equations (6.38), (6.39), (6.40), and (6.41) were split into individual terms resulting in the following set of scaled equations.
(6.38) is replaced by

$$x_0 = 0.1 \tag{6.42}$$
$$x_1 = -10(x + f(x)) \tag{6.43}$$
$$x_2 = y + \frac{1}{2}x_1 \tag{6.44}$$
$$x = 3.12 \int x_2 \, dt + x_0 \tag{6.45}$$

where (6.42) represents the initial condition for the integration.
(6.39) is split into

$$y_1 = z - \frac{1}{8}y \tag{6.46}$$
$$y_2 = 1.25x + 2y_1 \tag{6.47}$$
$$y = 4 \int y_2 \, dt \tag{6.48}$$

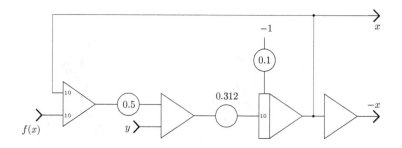

Fig. 6.60. Partial computer setup for equations (6.42), (6.43), (6.44), and (6.45)

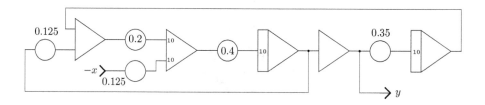

Fig. 6.61. Partial computer setup for equations (6.46), (6.47), (6.48), and (6.49)

while (6.40) becomes

$$z = -\int 3.5y \, dt. \tag{6.49}$$

The implementation of the central element, the CHUA diode, is quite costly as the implementation of each of the absolute value functions requires at least two diodes and an open amplifier. $f(x)$ is broken down and scaled like this:

$$f_1 = |0.7143x + 0.2857| \tag{6.50}$$
$$f_2 = |0.7143x - 0.2857| \tag{6.51}$$
$$f_3 = f_1 - f_2 \tag{6.52}$$
$$f(x) = -0.714x - 0.3003 f_3 \tag{6.53}$$

Deriving a computer setup from these equations is straightforward. Equations (6.42), (6.43), (6.44), and (6.45) are implemented as shown in figure 6.60, while equations (6.46), (6.47), (6.48), and (6.49)) are implemented as shown in figure 6.61.[126]

---

126 If the attractor does not appear, which can be caused by slightly uncalibrated integrators, increasing the value of the potentiometer set to 0.312 in figure 6.60 typically does the trick.

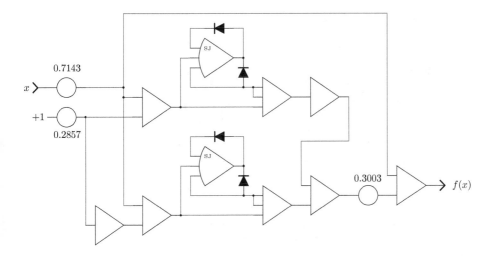

Fig. 6.62. Partial computer setup for equations (6.50), (6.51), (6.52), and (6.53).

It may be necessary to add small capacitors between the summing junction and the output of each of the two open amplifiers in the circuit for $f(x)$, if the setup shows signs of instability. Figure 6.62 shows the program for (6.50), (6.51), (6.52), and (6.53).

Figure 6.63 shows the implementation of this program on an early prototype of Analog Paradigm's Model-1 analog computer, which features a dedicated dual absolute value function module significantly simplifying the implementation of $f(x)$.

Depending on the time scale factors chosen for the integrators of this program, it is either possible to display a phase space plot of the Double Scroll Attractor on an oscilloscope or to plot it using a traditional pen plotter. Figure 6.64 shows a typical screenshot from a computation running with the time scale factors set to $k_0 = 10^3$.

6.19 Nonlinear chaos

Another surprisingly simple chaotic system is also based on an absolute value function as the central nonlinear element. This system has been described in [KIERS et al. 2003]. It is characterized by the following differential equation:

$$\dddot{x} = -\lambda \ddot{x} - \dot{x} + |x| - 1$$

A remarkable feature is that the behavior of this system is controlled by just one parameter, λ. The scaled computer program is shown in figure 6.65. If no absolute

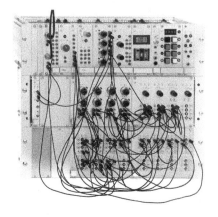
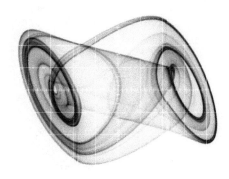

Fig. 6.63. Setup of the CHUA oscillator on an Analog Paradigm Model-1 analog computer prototype

Fig. 6.64. Phase space display of the Double Scroll attractor

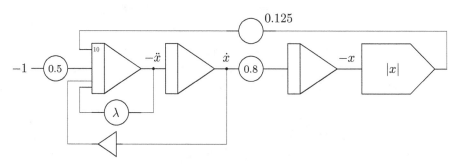

Fig. 6.65. Computer setup for nonlinear chaos

value function is readily available on the analog computer being used, one can be setup as shown in figure 5.20.

An example of a phase space plot based on $-\ddot{x}$ and $\dot{x}$ with $\lambda \approx 0.62$ is shown in figure 6.66.

6.20 Aizawa attractor

One of the prettiest chaotic attractors by far is the AIZAWA attractor,[127] which is described by the following three coupled differential equations

$$\dot{x} = x(z - \beta) - \delta y,$$

---

127 Cf. [COPE 2017].

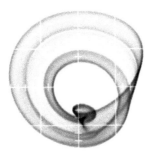

Fig. 6.66. Phase space plot of the chaotic attractor

$$\dot{y} = \delta x + y(z - \beta), \text{ and}$$
$$\dot{z} = \gamma + \alpha z - \frac{z^3}{3} - x^2 + \varepsilon z x^3$$

with the parameters $\alpha = 0.95$, $\beta = 0.7$, $\gamma = 0.65$,[128] $\delta = 3.5$, and $\varepsilon = 0.1$. A thorough numerical study of the behavior of this particular system can be found in [LANGFORD 1984].

Scaling these equations to ensure that no variable exceeds the machine units of ± 1 is straight-forward: The scaling factors corresponding to x, y, and z are $\lambda_x = \frac{1}{3}$, $\lambda_y = \frac{1}{4}$, and $\lambda_z = \frac{1}{2}$. The value $z - \beta$ is scaled by $\frac{2}{5}$.

The resulting computer setup is rather convoluted as shown in figure 6.67. In total three integrators, four summers, seven multipliers, and eleven coefficient potentiometers are required for this program. Figure 6.68 shows the x, z phase space plot of the AIZAWA attractor – an exceptionally beautiful structure. To achieve this two of the parameters resulting from the scaling, marked by * and ** in figure 6.67, were varied slightly to achieve a "nicer" picture. The values for these obtained by scaling are 0.95 and 0.27 respectively.

6.21 Nosé-Hoover oscillator

Another beautiful chaotic system is the NOSÉ-HOOVER oscillator, which is described in [SPROTT 2010, p. 95 et seq.]. This system is described by the following set of coupled differential equations:

$$\dot{x} = y$$
$$\dot{y} = yz - x$$
$$\dot{z} = 1 - y^2$$

---

[128] The original system has $\gamma = 0.6$, but 0.65 typically yields a better result.

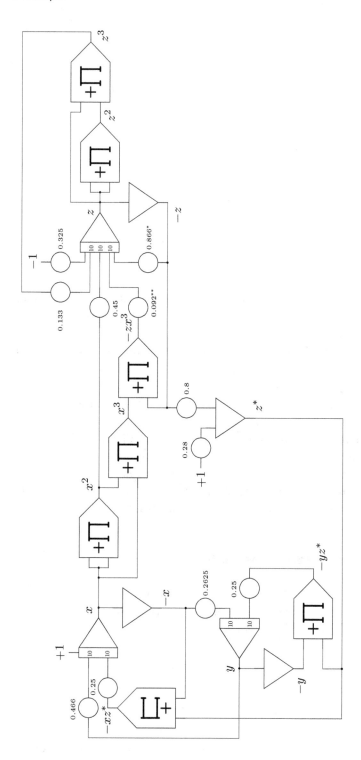

Fig. 6.67. Program for the AIZAWA attractor

6.21 Nosé-Hoover oscillator

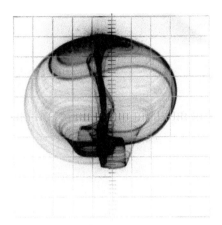

Fig. 6.68. The x, z phase space plot of the Aizawa attractor

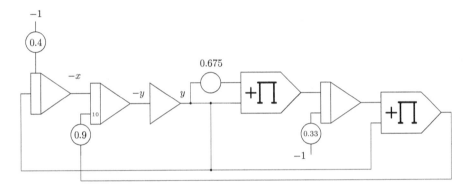

Fig. 6.69. Setup for the Nosé-Hoover oscillator

Scaling this system is surprisingly difficult but finally yields the following set of equations, which can be easily converted to the analog computer program shown in figure 6.69:

$$x = \int y \, dt$$
$$y = \int (-x + 9yz) \, dt$$
$$z = -\int \left(-\frac{1}{3} + 6.75y^2\right) dt$$

Figure 6.70 shows the mesmerizing phase space plot of the Nosé-Hoover oscillator.

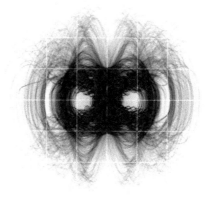

Fig. 6.70. Phase state plot of the NOSÉ-HOOVER oscillator

6.22 The SQ_M model

This section shows the implementation of the SQ_M model (cf. [SPROTT 2010, p. 68 et seq.]), an example of a simple three-dimensional chaotic flow with a quadratic nonlinearity at its center which is described by the following three coupled differential equations:

$$\dot{x} = -z$$
$$\dot{y} = -x^2 - y$$
$$\dot{z} = \alpha + \alpha x + y$$

Here, $\alpha = 1.7$ and the initial conditions are $x(0) = 1$, $y(0) = -0.8$, and $z(0) = 0$. To scale the system, three scale factors $\lambda_x = \lambda_z = \frac{1}{4}$ and $\lambda_y = \frac{1}{6}$ are introduced which in turn yield after collecting all resulting factors the following scaled system:

$$\dot{x} = -z \qquad (6.54)$$
$$\dot{y} = -2.666x^2 - y \qquad (6.55)$$
$$\dot{z} = \frac{\alpha}{4} + \alpha x + 0.15y. \qquad (6.56)$$

Thanks to the constant term $\frac{\alpha}{4}$, the initial conditions mentioned in the original system can be safely ignored as it will enter its chaotic oscillation quickly even without any explicit initial conditions.

The analog computer setup can be derived directly from the scaled equations (6.54), (6.55) and (6.56) as shown in figure 6.71. This program is ideally suited to be implemented on *THE ANALOG THING* as shown in figure 6.72. Figure 6.73 shows a typical phase space plot of x vs. y which was caputured using a

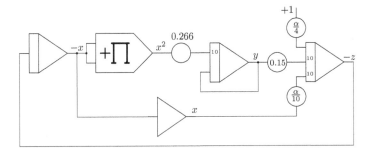

Fig. 6.71. Analog computer setup for the SQ_M model

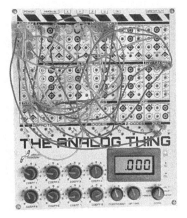
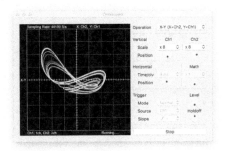

Fig. 6.72. Implementation of the program shown in figure 6.71

Fig. 6.73. xy phase space plot of the SQ_M system

USB-soundcard[129] with stereo line in and the software *Oscilloppoi*[130] running on a Mac.

6.23 The Duffing oscillator

A DUFFING *equation* describes any oscillator featuring a cubic stiffness term, i. e., a nonlinear elasiticity, such as

$$\ddot{x} + \delta\dot{x} + \alpha x + \beta x^3 = 0.$$

---

[129] It should be noted that most soundcards have AC coupled inputs, so any DC component present in a signal is removed. Typically this approach is only viable with the analog computer running in repetitive mode.
[130] See https://anikikobo.com/software/oscilloppoi/index_en.html.

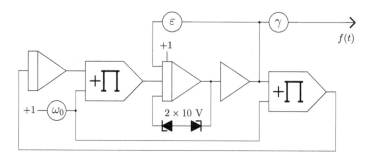

Fig. 6.74. Forcing function $f(t)$

It is named after GEORG WILHELM CHRISTIAN CASPAR DUFFING (1861 – 1944), a German engineer and inventor who wrote a seminal work on forced oscillations in 1918 (cf. [DUFFING 1918]).

This oscillator exhibits a nice chaotic behavior when driven by an external forcing function, i.e., in the form of

$$\ddot{x} + \delta\dot{x} + \alpha x + \beta x^3 = \gamma \cos(\omega_0 t).$$

Thorough theoretical treatments of this oscillator have been done and may be found in [KORSCH et al. 2008], [HÖHLER 1988], and [CHANG 2017]. Apart from its interesting mathematical properties it is a nice dynamic system which can be implemented on an analog computer without complicated scaling. It invites to playing with its parameters to achieve all kinds of nice phase space plots.

Figure 6.74 shows the implementation of the forcing function $f(t) = \gamma \cos(\omega_0 t)$. To achieve a minimum of harmonic distortion, the two amplitude limiting Zener diodes are connected to an integrator input with weight 1 instead of the summing junction. ε introduces a tiny positive feedback signal to avoid a decreasing amplitude, while γ controls the amplitude of the output signal.

Figure 6.75 shows the straight-forward implementation of the DUFFING oscillator. β was scaled down by a factor of $\frac{1}{10}$ to allow for values of β up to 10. A good initial parameter set to start with experiments is defined by $\alpha = 1$, $\beta = 5$, $\gamma = 1$, $\delta = 0.02$, $\omega_0 = 1$, and ε as small as possible.

Varying the parameters manually shows the complex behavior of this forced oscillator which even shows chaotic characteristics in certain cases. The effect of ω_0 is quite distinct. Two typical x, $-\dot{x}$ phase space plots are shown in figure 6.76. To get a wider range for γ, the output $f(t)$ may also be connected to an integrator input with weight 10 on the DUFFING oscillator sub-circuit.

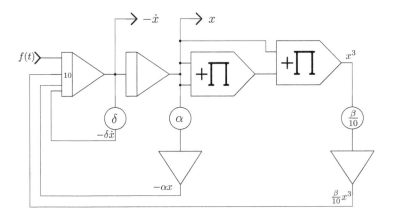

Fig. 6.75. DUFFING oscillator

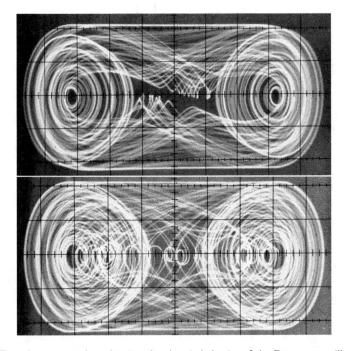

Fig. 6.76. Two phase space plots showing the chaotic behavior of the DUFFING oscillator

6.24 Rotating spiral

The following example displays a rotating spiral on an oscilloscope and is a real eye-catcher. The three-dimensional spiral is mapped to the two-dimensional oscilloscope screen using a circuit like the one shown in figure 6.54 in section 6.15. Since continuous rotation is required, the two potentiometers giving the values

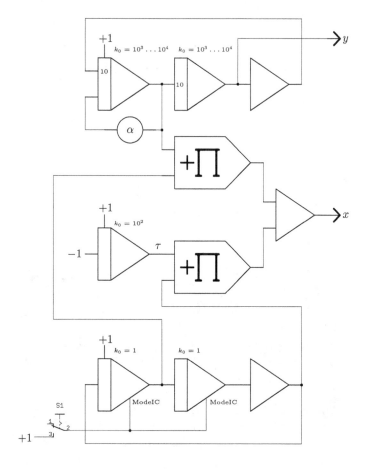

Fig. 6.77. Computer setup for the rotating spiral

$\sin(\varphi)$ and $\cos(\varphi)$ are replaced by a simple harmonic oscillator yielding a $\sin(\omega t)$ and $\cos(\omega t)$ signal pair.[131]

Figure 6.77 shows the overall computer setup. At the top is a harmonic oscillator running at high speed with a time scale factor of $k_0 = 10^3$ or, even better, $k_0 = 10^4$. The potentiometer α determines the amount of damping and thus the shape of the spiral. A three-dimensional spiral requires a third time-dependent variable τ, which is generated by the integrator in the middle of the schematic set to a time scale factor of $k_0 = 10^2$. These three integrators run in repetitive mode with an operating time of 20 ms.

The second oscillator at the bottom of figure 6.77 generates an undamped sine/cosine signal pair and is run in continuous mode, i.e., unaffected by the

---

131 See [LOTZ ASD].

global repetitive mode of operation. This is achieved by connecting the ModeIC inputs to +1 using a manual switch. If the switch is open,[132] these integrators are in IC-mode. With the switch closed both integrators are set to OP-mode. Their respective time scale factors are set to $k_0 = 1$.

The actual projection of the three-dimensional spiral onto two dimensions is done by the two multipliers followed by a summer shown in the middle of the schematic.

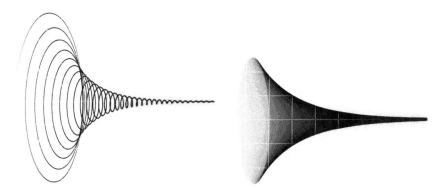

Fig. 6.78. Snapshots of rotating spirals with different time scale factors

Figure 6.78 shows two typical snapshots of the rotating spiral. In both cases the rotation had been stopped by placing the two oscillator integrators at the bottom of figure 6.77 into HALT-mode by means of their ModeOP-inputs (not shown in the setup). The picture on the left was taken with the time scale factors of the two topmost integrators set to $k_0 = 10^3$ while the right picture was obtained with $k_0 = 10^4$. If the integrators used do not offer this time scale factor directly, additional input resistors yielding a sufficiently large input weight can be connected to their SJ inputs. Input weights of up to 100 may be typically achieved by this technique.

6.25 Generating an Euler spiral

The first example in the beautiful book [HAVIL 2019, p. 1 et seq.] is the EULER *spiral*, which is a *clothoid*, a curve with curvature depending linearly on its arc length. Clothoids play a significant role in road and railway track construction. Imagine driving along a straight road approaching a curve. If the straight part of

---

[132] This applies to an Analog Paradigm Model-1 analog computer.

the road was directly connected to the arc of a circular curve, one would have to instantaneously change from a steering angle of zero to a non-zero steering angle, which is not just impracticable at low speeds but outright dangerous at higher velocities. Therefore, a safe road curve has a curvature that starts at zero, gently rises to its maximum and then falls back to zero again when the next straight road segment starts. These curves, which are encountered on many highways and railway tracks, can be implemented using clothoids.

Generally, a curve parameterized by some functions $x(t)$, $y(t)$ has a slope

$$m = \frac{dy}{dx} = \frac{\dot{y}}{\dot{x}}, \tag{6.57}$$

an arc length

$$l = \int_{t_0}^{t_1} \sqrt{\dot{x}^2 + \dot{y}^2}\, dt, \tag{6.58}$$

and a curvature

$$\kappa = \frac{\dot{x}\ddot{y} - \dot{y}\ddot{x}}{\sqrt{(\dot{x}^2 + \dot{y}^2)^3}}. \tag{6.59}$$

Using the general parameterization

$$x(t) = \int_0^T \cos(f(t))\, dt \text{ and } y(t) = \int_0^T \sin(f(t))\, dt$$

yields the following time derivatives:

$$\dot{x} = \cos(f(t)) \qquad\qquad \dot{y} = \sin(f(t))$$
$$\ddot{x} = -\dot{f}\sin(f(t)) \qquad\qquad \ddot{y} = \dot{f}\cos(f(t))$$

Inserting these into (6.57), (6.58), and (6.59) yields

$$m = \frac{\sin(f(t))}{\cos(f(t))} = \tan(f(t)),$$

$$l = \int_0^T \sqrt{\cos^2(f(t)) + \sin^2(f(t))}\, dt = \int_0^T dt = T, \text{ and}$$

$$\kappa = \frac{\dot{f}\left(\cos^2(f(t)) + \sin^2(f(t))\right)}{\sqrt{\left(\cos^2(f(t)) + \sin^2(f(t))\right)^3}} = \dot{f}.$$

The EULER spiral satisfies $\kappa = t$, i.e., $\dot{f} = t$ yielding $f(t) = \frac{t^2}{2}$ from which its parameterization

$$x(t) = \int_0^T \cos\left(\frac{t^2}{2}\right) dt \text{ and } y(t) = \int_0^T \sin\left(\frac{t^2}{2}\right) dt$$

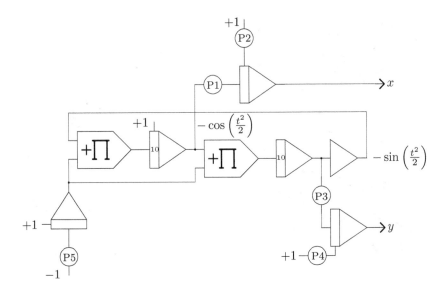

Fig. 6.79. Analog computer program for generating an EULER spiral

follows. This looks pretty straightforward to implement on an analog computer, but classic function generators for sin(φ) and cos(φ) have a limited interval for φ, typically $[-\frac{\pi}{2}, \frac{\pi}{2}]$. If more than a very short segment of the EULER spiral is to be generated by an analog computer, a trick must be employed. The idea is to use a quadrature generator as described in section 4.2. which expects $\dot{\varphi}$ instead of φ as its input. Such a generator, basically consisting of two integrators, two multipliers (to introduce $\dot{\varphi}$ into the underlying DEQ), and an inverter, is at the heart of the program shown in figure 6.79.

The quadrature output signals $\cos\left(\frac{t^2}{2}\right)$ and $\sin\left(\frac{t^2}{2}\right)$ are fed to integrators yielding $x(t)$ and $y(t)$, while $\dot{\varphi}$ is obtained by yet another integrator generating a linear ramp from -1 to $+1$. Figures 6.80 and 6.81 show the actual implementation of the program on *THE ANALOG THING* and a typical curve display as seen on an oscilloscope. The analog computer was run in repetitive mode with the operation time set so that the output of the integrator yielding $\dot{\varphi}$ ran from -1 to $+1$, which corresponds to roughly 20 ms. The overall parameter set is listed in table 6.1.

The growth of the EULER spiral can be seen on the oscilloscope by slowly increasing the OP-time. Ideally, all integrators are run with the highest time scale factor k_0 selected.[133]

---

[133] On THE ANALOG THING this corresponds to $k_0 = 10^3$.

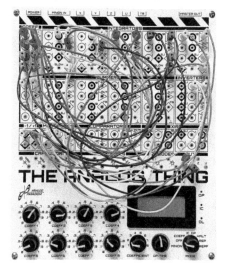 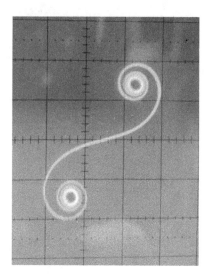

Fig. 6.80. Setup for generating an EULER spiral on THE ANALOG THING

Fig. 6.81. Typical output of the EULER spiral program

Coefficient	Description	Typical value
P1	x scaling	0.6
P2	x shift	0.87
P3	y scaling	0.6
P4	y shift	0.75
P5	ramp slope	0.1

Table 6.1. Parameters for the EULER spiral program

6.26 Hindmarsh-Rose model

From humble beginnings in the early 20th century the behavior of neurons has been described by increasingly realistic mathematical models. The very first of these models, *integrate-and-fire*, is due to LOUIS LAPICQUE, who developed it in 1907. An improved model was developed in the early 1960s by RICHARD FITZHUGH and J. NAGUMO and is described by the two coupled differential equations

$$\dot{v} = v - \frac{v^3}{3} - w + I_{\text{ext}} \text{ and}$$
$$\tau\dot{w} = v + a + -bw.$$

This model basically describes a relaxation oscillator[134] controlled by an external stimulus I_{ext} and is basically equivalent to the VAN DER POL equation

$$\ddot{y} + \mu\left(y^2 - 1\right)\dot{y} + y = 0.\text{[135]}$$

A much more recent model was suggested HINDMARSH and ROSE[136] and consists of three coupled differential equations

$$\dot{x} = -ax^3 + bx^2 + y - z + I_{\text{ext}} \qquad (6.60)$$
$$\dot{y} = -dx^2 + c - y \qquad (6.61)$$
$$\dot{z} = r(s(x - x_r) - z) \qquad (6.62)$$

with the parameters $a = 1$, $b = 3$, $c = 1$, $d = 5$, $r = 10^{-3}$, $s = 4$, $x_r = -\frac{8}{5}$, and initial conditions of 2 for all three integrators involved.

A quick numerical simulation, shown in figure 6.82, demonstrates that this system must be scaled before being implemented on an analog computer. Scaling is, as always, most easily done manually by first applying proper scaling factors to the variables x, y, and z. If, e.g., z is to be scaled by a factor of $\frac{1}{2}$, every input of the integrator yielding $-z$ must be scaled down by that factor. To compensate for this, z must be scaled up by a corresponding factor of 2 at all inputs of computing elements using this variable, etc. In the end, these various scaling factors tend to cancel out in most cases, so that typically only a few additional potentiometers are required for a scaled analog computer program compared to the unscaled equations.

The resulting scaled computer setup is shown in figure 6.83. The scaled parameters are $a^* = 4$, $b^* = 6$, $c^* = 0.066$, $d^* = 1.333$, $r = 10^{-3}$, $s = 4$, and $x_r^* = 0.8$ with initial conditions of ± 1 accordingly. Setting very small values such as r, rs, and rsx_r^* with manual potentiometers is not feasible. There are two techniques to overcome this problem: Use two coefficient potentiometers in series to allow for very small values or changing the time scale factor of the integrator to which these are connected.

In this example the latter approach was chosen: The time scale factor of the integrator yielding $-z$ was reduced to $k_0 = 10$. This requires all inputs to be scaled up by a factor of 10^2 to get the overall time scale factor of $k_0 = 10^3$. The advantage of this approach is that instead of a tiny value $rs = 0.004$ a much more feasible value of 0.4 is now required.

---

[134] In contrast to a harmonic oscillator which is typically based on an amplifier with suitable feedback, running in resonance mode, a relaxation oscillator switches abruptly between charge and discharge mode and thus yields non-harmonic output signals.
[135] See section 6.6.
[136] See [HINDMARSH et al. 1982] and [HINDMARSH et al. 1984].

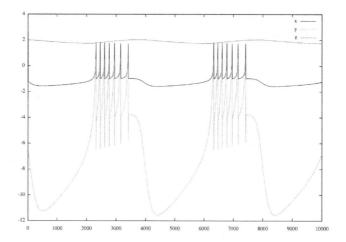

Fig. 6.82. Numerical simulation of the three coupled differential equations (6.60), (6.61), and (6.62)

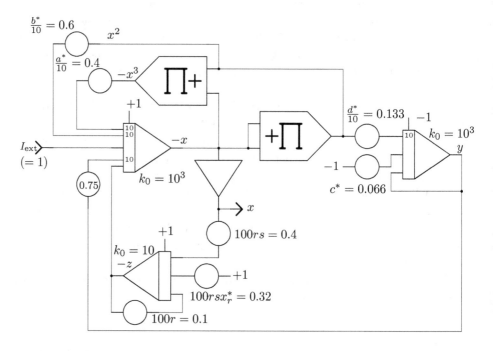

Fig. 6.83. Scaled analog computer setup for the HINDMARSH-ROSE model

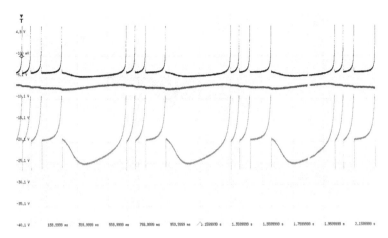

Fig. 6.84. Typical result of an analog spiking neuron simulation

Figure 6.84 shows a typical result obtained on a digital oscilloscope with this analog computer circuit and $I_{\text{ext}} = 1$.

6.27 Simulating the flight of a glider

In the early years of the 20th century, the English engineer and polymath FREDERICK WILLIAM LANCHESTER[137] discovered the phenomenon of *phugoid oscillation*, which describes a peculiar motion of an aircraft, in which it follows a sinusiodal flight path with respect to height over ground, pitching up and down and thus climbing and descending repeatedly instead of remaining in level flight.[138] A detailed description of this phenomenon can be found in [SIMANCA et al. 2002, p. 3:1 et seq.] on which the following derivation is based.

Figure 6.85[139] shows the basic glider aircraft considered here. φ is the angle between the centerline of the glider and the horizontal axis, while drag and lift are proportional to the square of the glider's velocity v. Introducing a drag coefficient R, the drag will be Rv^2 while lift will be considered to be equal to v^2 in the following.

Summing up the forces acting on the airplane yields

$$m\dot{v} = -mg\sin(\varphi) - Rv^2. \tag{6.63}$$

---

[137] 23.10.1868 – 08.03.1946
[138] See [LANCHESTER 1908].
[139] Cf. [SIMANCA et al. 2002, p. 3:2].

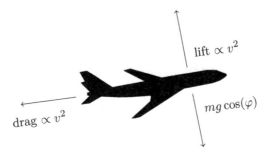

Fig. 6.85. Basic glider airplane

When the angle of attack becomes negative, the sine term will also become negative, yielding a positive mg-term on the right hand side, accelerating the airplane. A positive angle of attack will accordingly decelerate the airplane. Dividing both sides by m and changing R to R^* to absorb the $1/m$ term and setting the gravitational acceleration $g := 1$ to simplify things even further gives

$$\dot{v} = -\sin(\varphi) - R^* v^2. \tag{6.64}$$

The centripetal force acting on the airplane is

$$F_z = \frac{mv^2}{r}.$$

Since

$$\dot{\varphi} = \frac{v}{r}$$

this can be rewritten as

$$F_z = mv\dot{\varphi},$$

which must be equal to the sum of lift Lv^2 with some lift coefficient L and the downward force:

$$mv\dot{\varphi} = Lv^2 - mg\cos(\varphi)$$

Dividing by m, introducing a scaled lift coefficient L^* as before, and solving for varphi yields

$$\dot{\varphi} = L^* v - \frac{\cos(\varphi)}{v}. \tag{6.65}$$

The analog computer implementation of this problem is based on equations (6.64) and (6.65).

To display the flightpath of this glider, the required x, y-coordinate tuple can be generated by integrating over

$$\dot{x} = v\cos(\varphi) \text{ and}$$
$$\dot{y} = v\sin(\varphi).$$

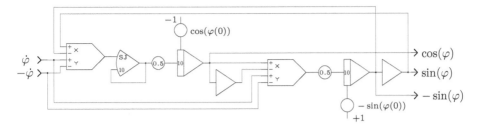

Fig. 6.86. Sine/cosine circuit

The parameters of this simulation are:

$v(0)$: Initial velocity of the glider.
R^*: Drag coefficient of the glider.
L^*: Lift coefficient, assumed to be 1 here.
$\varphi(0)$: Initial angle of attack.[140]

To make a (very long) story short, scaling this problem is pretty involved and may be left to the interested reader.[141]

This problem was solved on a historic Telefunken RA 770 analog computer.[142] Since this machine uses quarter square multipliers, all multipliers need their input values with both positive and negative signs unlike modern multipliers, thus cluttering the schematic a bit. Some of this machine's multipliers also require a dedicated buffer amplifier with a feedback of 10 instead of 1, which is pretty unusual from today's perspective. Accordingly, the following schematics show some technical detail which was typical for analog computer setups in the 1960s.

The first subcircuit is the generation of $\pm\sin(\varphi)$ and $\cos(\varphi)$ based on $\dot\varphi$ as input shown in figure 6.86. There is nothing special about this circuit. Since the simulation will typically be run in repetitive mode with rather high values of k_0, no amplitude stabilization is required here.

Figure 6.87 shows the partial computer setup yielding $\pm v$ and v^2. Note the input weights of 10 in this subcircuit. These were determined by manual scaling and yield a greater sensitivity of the glider to R^* and the gravitational pull.

---

[140] Since $\sin(\varphi)$ and $\cos(\varphi)$ are derived based on $\dot\varphi$ instead of φ since φ is not easily restricted to a fixed interval, setting $\varphi(0)$ requires the initial conditions of two integrators to be set to $\cos(\varphi(0))$ and $\sin(\varphi(0))$, respectively.
[141] This is the revenge for all the books the author read during his university days, which usually left unpleasant tasks like this to the reader...
[142] This particular model is without much debate the top model of Telefunken's analog computer family and was introduced in 1966 and built until 1975. It was often part of a hybrid computer installation. Not many of these machines are known to have survived.

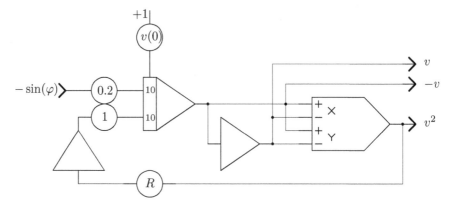

Fig. 6.87. Computing $\pm v$ and v^2, the scale factor 1 on the lower input of the integrator should be $\frac{1}{2}$ if scaled "by the book", but setting this parameter to 1 yields more sensitivity with respect to R

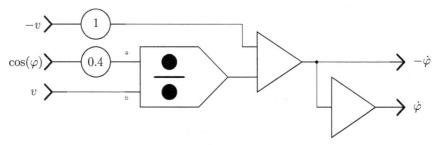

Fig. 6.88. Computing $\pm\dot\varphi$

Deriving $\pm\dot\varphi$ requires a divider. Typically, division on an analog computer should be avoided for two reasons: Scaling quickly gets pretty complicated with division circuits and division circuits based on an open amplifier with a multiplier in the feedback loop tend to be unstable making things even worse. A small external capacitor between the summer output and its summing junction will prevent such oscillations. In this example, the division could not be avoided. The actual implementation of the division circuit on the RA 770 is pretty involved and isn't shown in figure 6.88. Instead only the abstract symbol for a divider is used. Deriving the x and y coordinates of the glider, as shown in figure 6.89, is straight-forward since v, $\sin(\varphi)$, and $\cos(\varphi)$ are known already.

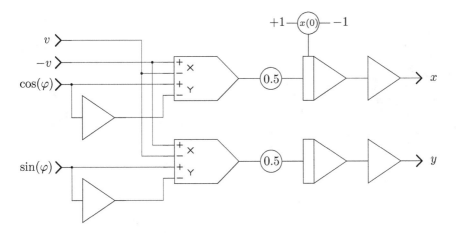

Fig. 6.89. Computing x and y of the glider

Parameter	Value
$-\sin(\varphi(0))$	0.707
$\cos(\varphi(0))$	0.707
$v(0)$	0.79
R^*	0.04
$-g$	0.107

Table 6.2. Parameters for the glider simulation

Figure 6.90 shows the actual setup on a Telefunken RA 770.[143] All in all this simulation requires seven summers, five integrators, 12 potentiometers,[144] one "free" potentiometer, five multipliers, and one divider.[145]

It is quite fun to play with the two parameters $v(0)$ and R. Figure 6.91 shows a typical simulation result. The simulation was run at high speed in repetitive mode so that a flicker free oscilloscope display could be achieved. This particular result was obtained with the parameters as shown in table 6.2.

The glider starts at $y = 0$ at the right and is thrown at a rather steep angle to the right. As v is pretty small, it goes into a dive, turns its direction to the left

---

[143] The little black boxes dangling from the patch panel are just distributors with six interconnected jacks. These are often necessary as the special Telefunken patch cables cannot be stacked together like banana plugs.
[144] Two of those could be eliminated as they have been set to 1 during the scaling process.
[145] Please note that the inverters required by the quarter square multipliers of the RA 770 have not been included in that list since these are only required due to the structure of this particular analog computer.

Fig. 6.90. Problem setup on a Telefunken RA 770

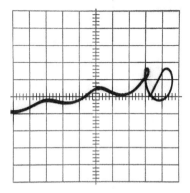

Fig. 6.91. typical simulation result

and gains velocity and thus lift which yields height. It makes a tiny loop before it slowly wiggles down.

6.28 Flow around an airfoil

The program described in this section simulates the flow of air around a special type of airfoil, a JOUKOWSKY *airfoil*.[146] The basic idea is to simulate the flow of air around an infinitely long rotating cylinder which is then transformed together with its surrounding airflow into the shape of an airfoil by a suitable *conformal mapping*.[147]

Spinning objects moving through air, such as golf balls or rotating cylinders, experience a force which is due to the effect of their rotation in conjunction with the surrounding medium. This is known as MAGNUS *effect* after HEINRICH GUSTAV MAGNUS, a German physicist who gave the first comprehensible explanation of this phenomenon. The airflow around a rotating cylinder will be simulated below using the following assumptions:

– The cylinder with circular cross section is infinitely long.

---

[146] Named after NIKOLAY YEGOROVICH JOUKOWSKY.
[147] This section is mainly based on [LOTZ AAB]. The author would like to thank Mr. LOTZ for several enlightening and interesting discussions regarding conformal mappings and their implementation on analog computers. Other useful sources are [REUTTER et al. 1968] and [SYDOW 1964, p. 123].

- The airflow is always perpendicular to the cylinder axis. Furthermore, it is stationary and conservative, i. e.,

$$\dot{v} = 0 \text{ and } \dot{A} = 0$$

with v and A denoting the velocity and area of the airflow.
- The air is assumed to be incompressible and frictionless.

This airflow is a *potential flow* and its velocity field can thus be described as the gradient of a complex function $f(z)$ called *velocity potential*. The function

$$f(z) = v(0) \left(z e^{i\varphi} + \frac{r^2 e^{-i\varphi}}{z} \right) - i\frac{\Gamma}{2\pi} \log(z)$$

will be used in this example. Here r denotes the radius of the cylinder around which the air flows. $v(0)$ is the initial velocity of the air flow, φ the angle of attack, and Γ the circulatory component.

First a circuit to generate the circumference of the rotating cylinder, i.e., a circle, is needed. This is implemented as always by solving the differential equation $\ddot{y} = -y$ as shown in figure 4.11 in section 4.2. The radius is set to about 0.4 by means of the parameter a^*, which controls the negative feedback, and the initial condition a of the first integrator. Both of these must match to avoid artefacts at the start of a simulation run. As before, the coefficient α provides a very small positive feedback. It can be omitted if the radius remains constant during a prolonged run. If required, it should be set to about 0.005, otherwise the sine/cosine signals will be distorted. The time scale factors of the two integrators should be set at least to $k_0 = 10^3$.

The program for the complex velocity potential shown in figure 6.92 is much more complicated. This circuit expects three parameters/input values:

- The angle of attack, $\tan(\varphi)$, which can be set manually by the free potentiometer shown in the upper left (if no free potentiometer is available, this parameter can be restricted to either positive or negative values only).
- A value τ varying linearly from -1 to $+1$ during one computer run. This represents the x-component of an air particle moving from left to right (the circuit actually requires $-\tau$).
- Finally, the input denoted $y(0)$ represents the distance of the flow line from the lower/upper surface of the airfoil. This value can be set either manually by a potentiometer connected to $+1$ or -1 or can be generated automatically, as shown below.

The small capacitors (ca. 47 pF to 68 pF) used in conjunction with the open amplifiers are required to stabilize the circuit and prevent it from oscillating. The

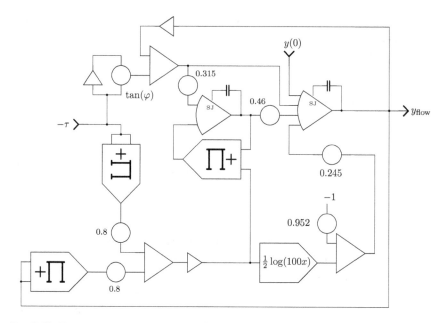

Fig. 6.92. Flowlines

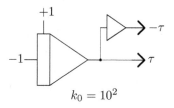

Fig. 6.93. Circuit yielding $\pm\tau$

function $\frac{1}{2}\log(100x)$ can be either generated by a diode function generator, by a TAYLOR approximation, or by the circuit described in appendix E.

Generating the time varying τ is easy, as shown in figure 6.93: One integrator (time scale factor set to $k_0 = 10^2$) and a summer suffice to yield both $-\tau$ and τ, which will be used later.

As nice as it is to control the vertical (start) position of a flow line manually, it is advisable to generate a time-dependent triangle signal y as shown in figure 5.15 in section 5.6. Since the computer is run in repetitive mode with an OP-time of 20 ms to get a flicker free picture, the integrator used to generate this signal must be controlled externally. On an Analog Paradigm Model-1 this can be achieved by patching the IC-input of one of the two rightmost integrators of an INT4 module to $+1$. The actual range of y is controlled by an additional coefficient potentiometer at the output of the triangle signal generator.

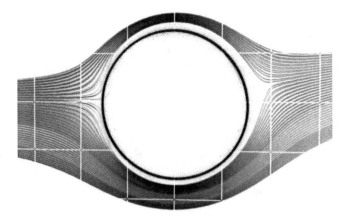

Fig. 6.94. Flow around rotating cylinder

Putting all of this together and using a display with two independent (x,y)-inputs the airflow around a rotating cylinder can be visualized as shown in figure 6.94.

Using a conformal or *angle-preserving* mapping it is now possible to transform the cylinder cross section into the shape of an airfoil. Applying the same mapping to the flow lines around the cylinder yields the corresponding flow lines around that particular airfoil.

Conformal mappings are functions generally defined on the complex plane $f : \mathbb{C} \to \mathbb{C}$ which locally preserve angles. One particularly demonstrative function which will be used here is the KUTTA-JOUKOWSKY *transform*

$$z = \zeta + \frac{1}{\zeta}$$

developed by MARTIN WILHELM KUTTA and NIKOLAI JOUKOWSKY. This function maps a (unit) circle into the shape of a special type of airfoil. A noteworthy property of this class of airfoils is a cusp at their trailing edge.

The actual conformal mapping as implemented is based on the transform

$$f(z) = (z - z_*) + \frac{\eta}{z - z_*},$$

which is split into its real and imaginary parts yielding

$$u(x(t), y(t)) = (x(t) - x_*) + \frac{\eta^2 (x(t) - x_*)}{(x(t) - x_*)^2 + (y(t) - y_*)^2} \text{ and}$$

$$v(x(t), y(t)) = (y(t) - y_*) - \frac{\eta^2 (y(t) - y_*)}{(x(t) - x_*)^2 + (y(t) - y_*)^2}.$$

The parameters x_* and y_* define the shape of the resulting airfoil. Good values to start exploration with are $x_* = 0.045$ and $y_* = 0.052$.[148] Figure 6.95 shows the computer setup for the conformal mapping.

The two open amplifiers used to implement the required division operations should have small capacitors connected between their respective outputs and summing junctions, as noted in section 5.2.2 and in the glider example before.

Since the conformal mapping circuit is required twice – to transform the circle into a JOUKOWSKY airfoil and to transform the airflow around the cylinder into the airflow around that airfoil – the inputs $x(t)$ and $y(t)$ of the circuit shown in figure 6.95 are rapidly switched between the outputs of the circuit generating $r\sin(\omega t)$ and $r\cos(\omega t)$ and the circuit yielding the coordinates of an air particle flowing around the hypothetical cylinder.

This can be implemented easily by means of two electronic switches as shown in figure 6.96. The switches are controlled in such a way that they change position every other repetitive cycle of the computer. This requires a toggle flip-flop driven by the operate control line of the computer.

The overall computer setup is quite complex and requires a large complement of computing elements, the majority of which are summers and multipliers. Figure 6.97 shows the resulting simulated airflow around an airfoil.

6.29 Heat transfer

This example deals with heat flow in a cable – a problem that is described by a partial differential equation, i.e., a differential equation containing differentials with respect to more than one variable. Such problems are frequently found in physics, engineering, and other areas and cannot easily be solved directly by an analog-electronic analog computer, because such a machine only allows integration with respect to (machine) time.

A straightforward approach to treat partial differential equations on an analog computer is to approximate the differentials which are not with respect to time as differential quotients, thus discretizing the underlying space. Depending on the required resolution this approach will require many computing elements, as can be seen below.[149]

---

[148] To reliably set values as small as these it is often beneficial to use two coefficient potentiometers in series, the first set to a value like 0.1 for some prescaling.

[149] More general information on this topic can be found in [BRYANT et al. 1962, p. 37 et seq.] and [GILLILAND 1967, p. 14-1 et seq.] as well as in other standard texts. The following example is based on [Telefunken 1963] and [GILOI et al. 1963, p. 264 et seq.]. A vastly different approach is described in [ALBRECHT 1968] where linear partial differential equations are transformed into ordinary differential equations by means of a LAPLACE transform (see appendix A). Section 7.7

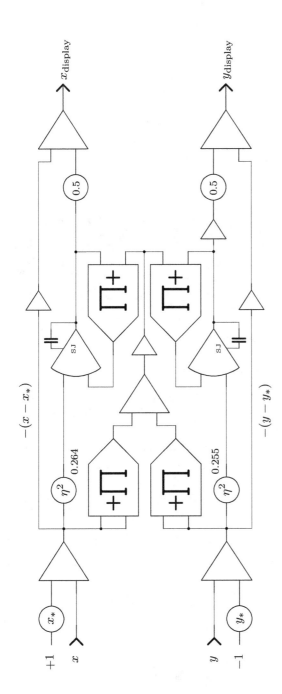

Fig. 6.95. Conformal mapping

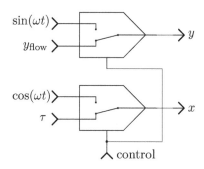

Fig. 6.96. Switching between profile and airflow

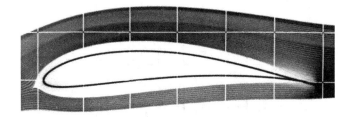

Fig. 6.97. Typical air flow around an airfoil

Figure 6.98 shows a cross section of a hypothetical cable consisting of a center conductor of radius r_c surrounded by an insulating material with an outer radius r_i. The current flowing through the center conductor heats it to a constant temperature T_0. The outside of the insulated cable is held at a fixed temperature 0. The aim of this simulation is to determine the heat distribution within the insulating material with respect to time t and radius r.

The general homogenous form of the *heat equation* has the form

$$\frac{\partial T}{\partial t} - a\nabla^2 T = 0 \qquad (6.66)$$

where a is a constant term describing the *thermal diffusivity* of the material. Generally, it is

$$a = \frac{k}{\rho c_\mathrm{p}}$$

---

shows another approach to the solution of partial differential equations on an analog/hybrid computer.

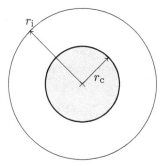

Fig. 6.98. Structure of the heated cable

with k denoting the *thermal conductivity* of the material. ρ is its *density*, and c_p is the *specific heat capacity*.

The temperature T within the material is a function of time t and a location vector $\vec{x}$, i.e., $T(\vec{x},t)$. The arguments of the function T are omitted as always to avoid unnecessary clutter in the equations.

∇^2 is the LAPLACE operator, which is often also denoted by Δ. It is

$$\nabla^2 T = \Delta T = \mathrm{div}(\mathrm{grad}(T)). \tag{6.67}$$

Due to the inherent rotational symmetry of the two-dimensional cross section of the cable shown in figure 6.98, a one-dimensional flat slice running from the center to the circumference is sufficient to describe the overall problem.

Since the cable in question contains a heat source, the general heat equation (6.66) has to be adapted as follows by adding another derivative term:

$$\frac{\partial T}{\partial t} = a\left(\frac{\partial^2 T}{\partial r^2} + \frac{\partial T}{r\partial r}\right) \tag{6.68}$$

This problem is further determined by the following boundary conditions as described before:

$$T(r,t) = \begin{cases} T_0 = \mathrm{const.} & \text{if } r \leq r_\mathrm{c} \\ 0 & \text{if } r > r_\mathrm{i} \end{cases}$$

Additionally, the following initial conditions hold assuming that the insulator was uniformly cooled to 0 at $t = 0$:

$$T(r,0) = \begin{cases} T_0 = \mathrm{const.} & \text{if } r \leq r_\mathrm{c} \\ 0 & \text{else} \end{cases} \tag{6.69}$$

In order to devise an analog computer program to solve (6.68) the insulator with thickness $r_\mathrm{i} - r_\mathrm{c}$ will be subdivided into $n \in \mathbb{N}$ slices of equal thickness Δr.[150]

---

[150] Here the Δ denotes a difference – not the LAPLACE operator!

Thus, the $T(r,t)$ are replaced by a finite number of $T_i(t)$ by approximating the differentials by differential quotients with respect to r yielding a number of coupled ordinary differential equations which can then be readily solved by an analog computer. This is based on the following approximations for first and second derivatives

$$\left.\frac{\partial f(x,t)}{\partial x}\right|_{x_i} \approx \frac{f_{i+1}(t) - f_{i-1}(t)}{2\Delta x} \quad \text{and} \tag{6.70}$$

$$\left.\frac{\partial^2 f(x,t)}{\partial x^2}\right|_{x_i} \approx \frac{f_{i+1}(t) - 2f_i(t) + f_{i-1}(t)}{(\Delta x)^2}. \tag{6.71}$$

Applying (6.70) and (6.71) to (6.68) yields the following general expression

$$\dot{T}_i = a\left(\frac{T_{i+1} - 2T_i + T_{i-1}}{(\Delta r)^2} + \frac{1}{r_\mathrm{c} + i\Delta r}\frac{T_{i+1} - T_{i-1}}{2\Delta r}\right), \tag{6.72}$$

which can be readily implemented on an analog computer since it only contains a derivative with respect to t.

In the following example the insulator will be divided into four equally wide slices. It will be further assumed that $r_\mathrm{i} = 5r_\mathrm{c}$ yielding

$$\Delta r = \frac{r_\mathrm{i} - r_\mathrm{c}}{4} = r_\mathrm{c}.$$

Accordingly, (6.72) can be rewritten as follows:

$$\dot{T}_i = a\left(\frac{T_{i+1} - 2T_i + T_{i-1}}{(\Delta r)^2} + \frac{T_{i+1} - T_{i-1}}{2(i+1)(\Delta r)^2}\right)$$

$$= \frac{a}{(\Delta r)^2}\left((T_{i+1} - 2T_i + T_{i-1}) + \frac{1}{2(i+1)}(T_{i+1} - T_{i-1})\right)$$

$$= \frac{a}{(\Delta r)^2}\left(\left(1 + \frac{1}{2(i+1)}\right)T_{i+1} - 2T_i + \left(1 - \frac{1}{2(i+1)}\right)T_{i-1}\right)$$

The terms $1 + \frac{1}{2(i+1)}$ will, of course, exceed 1, so some rough scaling is necessary. Introducing a scale factor λ such that

$$\frac{a}{\lambda(\Delta r)^2} = \frac{1}{2}$$

solves this problem and yields the following set of coupled ODEs:

$$T_0 := 1$$

$$\dot{T}_1 = \frac{1}{2}\left(\frac{5}{4}T_2 - 2T_1 + \frac{3}{4}T_0\right) = \frac{5}{8}T_2 - T_1 + \frac{3}{8}T_0$$

$$\dot{T}_2 = \frac{1}{2}\left(\frac{7}{6}T_3 - 2T_2 + \frac{5}{6}T_1\right) = \frac{7}{12}T_3 - T_2 + \frac{5}{12}T_1$$

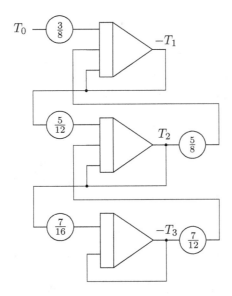

Fig. 6.99. Heat flow simulation

$$\dot{T}_3 = \frac{1}{2}\left(\frac{9}{8}T_4 - 2T_3 + \frac{7}{8}T_2\right) = \frac{9}{16}T_4 - T_3 + \frac{7}{16}T_2$$

$$T_4 := 0$$

A clever trick often used in cases like this is to invert the signs of every other equation thus saving inverters by taking the implicit sign inversion that every integrator causes into account. Since the rough discretization shown above yields only three coupled ODEs, the second one must be rewritten as

$$-\dot{T}_2 = -\frac{7}{12}T_3 + T_2 - \frac{5}{12}T_1. \tag{6.73}$$

The resulting program is shown in figure 6.99 – due to the deliberate sign reversal in (6.73) no additional inverters are required. In general, $n-1$ integrators are needed to solve such a partial differential equation in which the derivatives with respect to the variable other than time are discretized by splitting them into n slices. In order to minimize the error caused by this discretization the number of slices should be as large as possible; three integrators, as in this example, is typically not enough to obtain reasonable solutions.

6.30 Two-dimensional heat transfer

Using the same basic idea of discretizing a continuous space into a discrete grid of cells as in the previous example it is also possible to solve higher dimensional partial differential equations such as the two-dimensional heat equation[151]

$$\dot{u} = \alpha \nabla^2 u = \alpha \left(\frac{\partial^2 u}{\partial x^2} + \frac{\partial^2 u}{\partial y^2} \right). \tag{6.74}$$

α represents the diffusity of the two-dimensional medium and is here set to $\alpha = 1$ to simplify things. Discretizing (6.74) along the x- and y-axes yields

$$\dot{u}_{i,j} = \alpha \left(u_{i-1,j} + u_{i+1,j} + u_{i,j-1} + u_{i,j+1} - 4 u_{i,j} \right) + q_{i,j} \tag{6.75}$$

describing a single node in the grid. $q_{i,j}$ represents an optional heat source or sink at the node $u_{i,j}$. In the following example heat is only applied to or removed from node $u_{0,0}$, i.e., $q_{i,j} = 0 \, \forall i, j \neq 0$.

Typically it is desirable to have as many grid nodes as possible in a simulation to minimize the error caused by the above discretization process. Some classic analog computer installations, especially those in the chemical industry which were used for the design of fractionating columns, often featured many hundred integrators and sometimes in excess of a thousand coefficient potentiometers. Whenever possible, inherent symmetries of a problem should be exploited in order to conserve computing elements.

In the following case a 8×8 cm^2 plate, perfectly insulated on its top and bottom surfaces, is considered. The plate's edges are held at a fixed temperature T. Quadratic square plates like this exhibit symmetries not only along their x- and y-axes but also with respect to their diagonals as shown in figure 6.100. Accordingly, it is sufficient to take only one octant of the plate into account in the simulation, thus saving a considerable number of computing elements. Nevertheless, this comes at the cost that additional boundary conditions are required.

If only an octant instead of a quadrant of the plate as shown in figure 6.100 is implemented as a grid of computing nodes, the nodes along the diagonal and those along the x-axis require some attention as they have to be connected to their "mirror" neighbor nodes which replace the missing neighbor nodes in the adjacent octants. The nodes along the vertical right hand side of the octant are held at the fixed temperature T. The node in the center of the plate is denoted by $u_{0,0}$ and is the only node that allows heat to be applied or extracted.

According to equation (6.75), every integrator in the grid yielding $u_{i,j}$ requires five inputs: The outputs of its direct neighbors $u_{i-1,j}$, $u_{i+1,j}$, $u_{i,j-1}$, and $u_{i,j+1}$

---

[151] The author is deeply indebted to Dr. CHRIS GILES, who not only suggested this problem and that of appendix B, but also did the majority of the implementation.

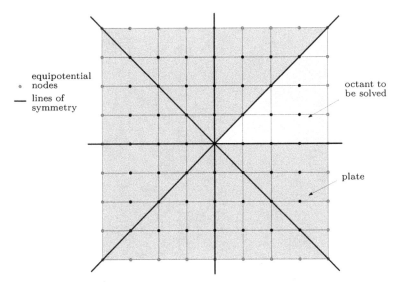

Fig. 6.100. The heat conducting plate

as well as its own inverted output yielding the $-4u_{i,j}$ term. Node $u_{0,0}$ requires an additional sixth input for $q_{0,0}$.

The various paralleled inputs of the elements along the diagonal should be clustered into single inputs with weight 10 and a coefficient potentiometer suitably set in the actual computer setup.

Figure 6.102 shows two typical results obtained by this setup. The input $q_{0,0}$ has been subject to a step impulse at the start of the computer run. The picture on the left shows this impulse in the top trace followed by the outputs of the nodes $u_{0,0}$, $u_{1,0}$, and $u_{2,0}$. The picture on the right only shows the outputs of the nodes along the diagonal, i.e., $u_{0,0}$, $u_{1,1}$, $u_{2,2}$, and $u_{2,2}$.[152]

Figure 6.103 again shows the impulse fed into node $u_{0,0}$ and the outputs of the following three nodes along the x-axis but this time with a fixed boundary temperature $T = 1$. It can be clearly seen that the nodes are "heated" by the boundary while the heat spike $q_{0,0}$ accelerated the heating initially.

6.31 Systems of linear equations

It may seem bizarre to use analog computers, which are usually associated with the solution of differential equations, to solve sets of linear equations. Such equations

---

[152] Note that the vertical sensitivity has been increased by a factor 2 with respect to the right pictures in order to obtain a reasonable output for the grid nodes further along the diagonal.

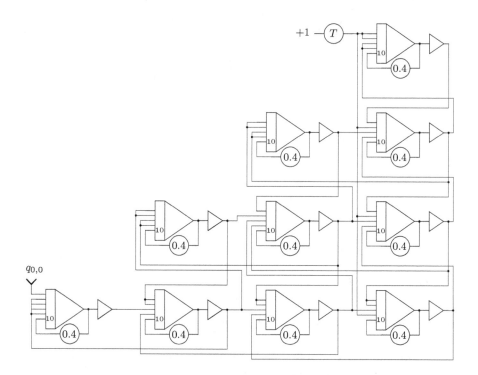

Fig. 6.101. Basic setup of the two-dimensional heat problem

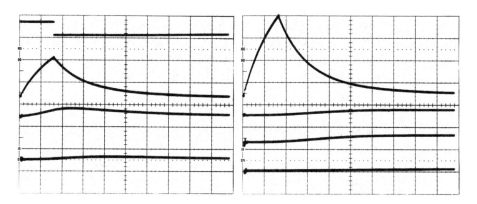

Fig. 6.102. Typical results with $T = 0$ at the plate boundary

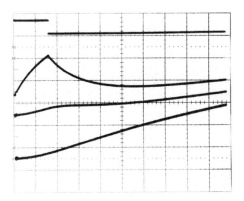

Fig. 6.103. Results along the x-axis with non-zero boundary temperature T

are typically solved numerically using an algorithmic approach such as the GAUSS-SEIDEL or GAUSS-JORDAN methods. Nevertheless, recent developments in artificial intelligence and other areas have spurred the idea of using an analog computer as a linear algebra accelerator for solving systems of linear equations.[153]

In the following $\mathbf{A}$, $\vec{b}$ and $\vec{x}$ are defined as follows:[154]

$$\mathbf{A} = \begin{pmatrix} a_{11} & a_{12} & \cdots & a_{1n} \\ \vdots & \vdots & \ddots & \vdots \\ a_{n1} & a_{n2} & \cdots & a_{nn} \end{pmatrix} \in \mathbb{R}^{n \times n}, \vec{b} = \begin{pmatrix} b_1 \\ \vdots \\ b_n \end{pmatrix} \in \mathbb{R}^n, \vec{x} = \begin{pmatrix} x_1 \\ \vdots \\ x_n \end{pmatrix} \in \mathbb{R}^n.$$

It is assumed that the matrix $\mathbf{A}$ is non-singular. The task at hand is now to solve a system of linear equations such as

$$\mathbf{A}\vec{x} = \vec{b} \tag{6.76}$$

for a given $\mathbf{A}$ and $\vec{b}$, i.e.,

$$\sum_{j=1}^{n} a_{ij} x_j - b_i = 0 \text{ with } 1 \leq i \leq n.$$

A direct approach would transform such a system of linear equations into a corresponding analog computer setup, which will be demonstrated using a simple example of two coupled equations:

$$a_{11} x_1 + a_{12} x_2 - b_1 = 0 \tag{6.77}$$

---

[153] Cf. [HUANG et al. 2017]. The work presented in this section has been done in collaboration with DIRK KILLAT, see [ULMANN et al. 2019].
[154] $\mathbf{A} \in \mathbb{C}^{n \times n}, \vec{b} \in \mathbb{C}^n$, and $\vec{x} \in \mathbb{C}^n$ are not ruled out but have to be split into their respective real and imaginary parts in order to apply the methods described here.

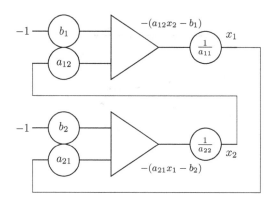

Fig. 6.104. Unsuitable direct approach of solving a system of linear equations on an analog computer

$$a_{21}x_1 + a_{22}x_2 - b_2 = 0 \tag{6.78}$$

Solving (6.77) and (6.78) for x_1 and x_2 respectively yields

$$x_1 = \frac{b_1 - a_{12}x_2}{a_{11}} \text{ and}$$

$$x_2 = \frac{b_2 - a_{21}x_1}{a_{22}},$$

which can be readily transformed into the analog computer setup shown in figure 6.104.

Although this direct approach looks elegant, it doesn't work! First, the coefficients $1/a_{ii}$ are hard to implement as an analog computer is always restricted to values in the interval $[-1; 1]$. Second, the main problem is that the setup shown contains *algebraic loops*, which are loops consisting of an even number of summers. Since every summer performs an implicit sign inversion an even number of such elements will yield the same sign at its output as at its input, resulting in a positive feedback loop, which will result in an inherently unstable, typically oscillating, circuit.

Accordingly, another approach, better suited to an analog computer, is required. The basic idea is to transform a system of linear equations such as (6.76) into a system of coupled differential equations. The solution vector $\vec{x}$ is then initialized with some initial value and an error vector $\vec{\varepsilon}$ is computed, which is then used to correct this initial guess of $\vec{x}$:

$$\mathbf{A}\vec{x} = \vec{\varepsilon} \text{ and}$$
$$\dot{\vec{x}} = -\vec{\varepsilon}$$

or, in a component-wise-fashion

$$-\dot{x}_i = \sum_{j=1}^{n} a_{ij} x_j - b_i, 1 \leq i \leq n \tag{6.79}$$

where $\dot{x}_i$ denotes the time derivative of x_i.

A set of equations like (6.79) can be directly transformed into an analog computer setup by applying the KELVIN feedback technique. Figure 6.105 shows this for a system with three unknowns, i.e.,

$$-\dot{x}_1 = a_{11} x_1 + a_{12} x_2 + a_{13} x_3 - b_1$$
$$-\dot{x}_2 = a_{21} x_1 + a_{22} x_2 + a_{23} x_3 - b_2$$
$$-\dot{x}_3 = a_{31} x_1 + a_{32} x_3 + a_{33} x_3 - b_3.$$

Although this approach looks elegant it still has a major drawback in that it often does not converge. Basically, the underlying system of coupled differential equations is stable when $\dot{x}_i = 0 \forall i$. This requires A to be symmetric and positive definite, i.e., all roots of the characteristic equation

$$|\mathbf{A} - \lambda \mathbf{I}| = 0 \tag{6.80}$$

with $\mathbf{I}$ denoting the $n \times n$ unit matrix must be greater than zero. Unfortunately, this is often not the case, making this direct approach unsuitable for most real applications.

Figure 6.106 shows the non-convergent behavior of this simple indirect approach applied to the following system of linear equations:

$$\begin{pmatrix} 0.8 & 0.5 & 0.3 \\ 0.1 & 0.6 & 0.8 \\ 0.2 & 0.9 & 0.4 \end{pmatrix} \begin{pmatrix} x_1 \\ x_2 \\ x_3 \end{pmatrix} = \begin{pmatrix} 0.8 \\ 0.7 \\ 0.3 \end{pmatrix} \tag{6.81}$$

A clever solution to this problem is to transform the original problem $\mathbf{A}\vec{x} = \vec{b}$ into a corresponding system of linear equations with a coefficient matrix that is always positive definite.[155] This could be easily achieved by multiplying (6.76) by the transposed matrix $\mathbf{A}^T$ from the left yielding

$$\mathbf{A}^T \mathbf{A} \vec{x} = \mathbf{A}^T \vec{b}. \tag{6.82}$$

The matrix $\mathbf{H} = \mathbf{A}^T \mathbf{A}$ is positive definite and Hermitian[156] so that this modified system of linear equations can be solved by the approach shown above. A

---

155 More information can be found in [FIFER 1961, p. 842 et seq.], [ADLER 1968, p. 281 et seq.], and [KOVACH et al. 1962].
156 A rather technical proof of this may be found in [FIFER 1961, p. 842 et seq.].

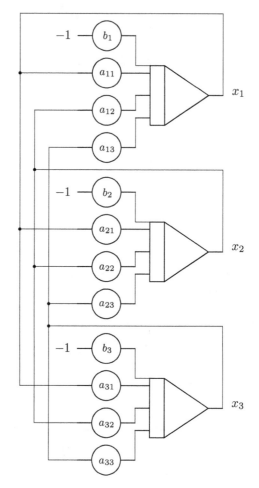

Fig. 6.105. Indirect approach to solving a system of linear equations on an analog computer

direct approach might convert (6.76) into (6.82) by means of a preliminary step performed on a digital computer.

Nevertheless, a purely analog implementation of this approach can be achieved by computing the error resulting from an initial setting of $\vec{x}$ by

$$\mathbf{A}^T \mathbf{A} \vec{x} - \mathbf{A}^T \vec{b} = \vec{\zeta}$$

and setting

$$\dot{\vec{x}} = -\vec{\zeta}$$

as before. Factoring out $\mathbf{A}^T$ yields

$$\mathbf{A}^T(\mathbf{A}\vec{x} - \vec{b}) = \mathbf{A}^T \vec{\varepsilon} = \vec{\zeta} = -\dot{\vec{x}}.$$

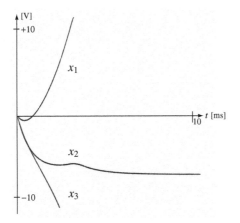

Fig. 6.106. Divergent behavior of the computer setup shown in figure 6.105

This can be now split into two sets of coupled equations:

$$\varepsilon = \sum_{j=1}^{n} a_{ij} x_j - b_i \qquad (6.83)$$

$$\dot{x}_i = -\sum_{j=1}^{n} a_{ji} \varepsilon_i \qquad (6.84)$$

with $1 \leq i \leq n$. These equations can now be transformed into an analog computer setup as shown in figure 6.107. Compared with the simple program shown in figure 6.104, this one requires twice the number of coefficient potentiometers (note the transposed indices) and summers. This seemingly profligate use of computing elements is nevertheless justified given the very good convergence behavior of this setup.

Figure 6.108 shows the complete setup using this advanced indirect method for solving the 3×3 system shown in equation 6.81 and the rapidly converging results are shown in figure 6.109. Not only is the rate of convergence fast, the results also match the solution obtained by using classic numerical approaches quite well:

$$\vec{x}_{\text{numerical}} = \begin{pmatrix} 0.8078 \\ -0.2852 \\ 0.9878 \end{pmatrix}, \vec{x}_{\text{analog}} = \begin{pmatrix} 0.8093 \\ -0.2877 \\ 0.9918 \end{pmatrix}$$

6.32 Human-in-the-loop

Analog computers are ideally suited for *human-in-the-loop* simulations where an operator interacts with a simulation such as a flight simulator, a nuclear power

194 — 6 Examples

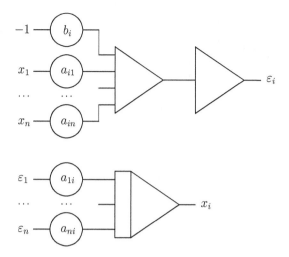

Fig. 6.107. Computer setup for equations (6.83) and (6.84)

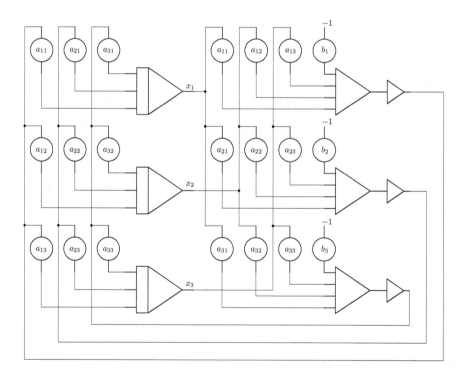

Fig. 6.108. Computer setup for a 3×3 system $\mathbf{A}\vec{x} = \vec{b}$

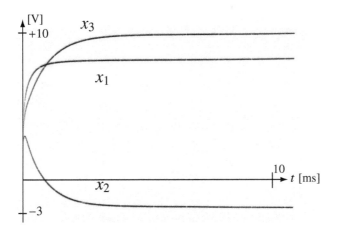

Fig. 6.109. Solution of (6.84) by the circuit shown in figure 6.108

plant simulator, etc. This section shows a simple example of this type of simulation. A user controls the rotational and translational movements of a spacecraft in empty space by firing rotational and translational thrusters using a joystick.[157]

First of all, a spacecraft-like figure is required for the display. Using a sine/cosine quadrature generator[158] a unit circle can be generated which is then suitably deformed using the circuit shown in figure 6.110 yielding the shape shown in figure 6.111.[159] The only important requirement here is that the spacecraft figure has a distinguishable front and back. The output of this subcircuit forms a time varying two-element vector

$$\vec{s} = \begin{pmatrix} s_x \\ s_y \end{pmatrix}$$

with $\lambda_x = 0.04$ and $\lambda_y = 0.05$ being reasonable values to give a suitably sized shape.

Next, suitable signals have to be derived from the joystick interface to control the angular velocity $\dot{\varphi}$ and the longitudinal acceleration a of the spacecraft. Figure 6.112 shows the corresponding computer setup with $-1 \leq j_x \leq 1$ and $-1 \leq j_y \leq 1$

---

[157] Cf. [ULMANN 2016] and see appendix G for a simple joystick interface.
[158] See figure 4.11 in section 4.2.
[159] Using a diode function generator instead of this simple setup can yield much a more sophisticated spacecraft shape.

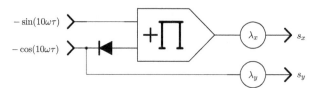
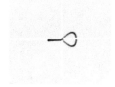

Fig. 6.110. Creating a spacecraft-like shape by deforming a circle

Fig. 6.111. Shape of the spacecraft as displayed on an x, y-display

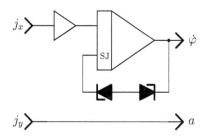

Fig. 6.112. Deriving $\dot{\varphi}$ and a from the joystick interface

representing the two output signals from the joystick.[160] The two ZENER-diodes in series, each with a ZENER-voltage of 10 V, limit the output signal of the integrator to avoid an overload condition.

Using the angular velocity $\dot{\varphi}$, the corresponding function values $\pm \sin(\varphi)$ and $\pm \cos(\varphi)$, which will be used to rotate the spacecraft, can be derived. Figure 6.113 shows the computer setup yielding these functions.

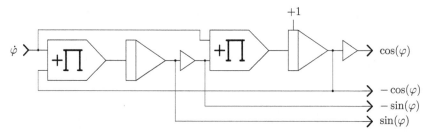

Fig. 6.113. Generating $\pm \sin(\varphi)$ and $\pm \cos(\varphi)$

---

160 Using comparators and electronic switches in conjunction with one integrator each, even cheap digital "retro gaming" joysticks may be used instead of an analog joystick to produce continuous voltage signals.

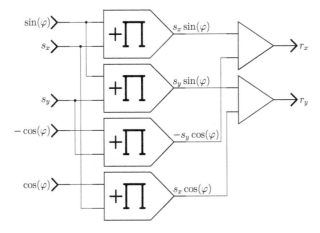

Fig. 6.114. Rotating the spacecraft shape

The actual rotation is then performed by computing

$$\vec{r} = \begin{pmatrix} -\sin(\varphi) & \cos(\varphi) \\ -\cos(\varphi) & -\sin(\varphi) \end{pmatrix} \vec{s}$$

as shown in figure 6.114 where $\vec{r}$ denotes the time-varying vector of the rotated spacecraft shape. The computer setup for implementing this rotation matrix is shown in figure 6.114. If the computer has *resolver* units, one of them can be used to implement the rotation instead of using four multipliers and two summers.[161]

This rotated spacecraft shape must now be able to move around on the display screen controlled by "firing" its translational thrusters which are assumed to work only along its longitudinal direction. The acceleration a from the joystick controller therefore must be integrated twice in a component-wise fashion taking the current angle of rotation φ into account. Figure 6.115 shows the subprogram for this translational movement.

In total, the overall program requires eight summers, nine integrators, six coefficient potentiometers, nine multipliers, several free diodes, an x, y-display, and an analog joystick.

---

[161] Basically, a resolver combines the required function generators, multipliers, and summers, which are required for operations like the transformation of polar coordinates into rectangular ones, and vice versa, or the rotation of coordinate systems.

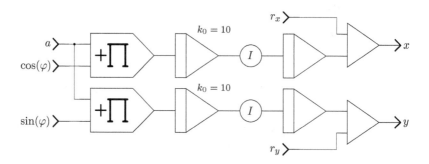

Fig. 6.115. Translational movement of the spacecraft

6.33 Inverted pendulum

This section describes the simulation of an inverted pendulum, i.e., a pendulum that is mounted on a cart capable of moving along the x-axis only and has its mass above its pivot point, as shown in figure 6.116. The aim is to keep this pendulum in its upright position by controlling a force F acting on the cart in order to move it from left to right and vice versa.

If such a pendulum were mounted on a stationary cart, its equation of motion would be the same as that for a mathematical pendulum, i.e.,

$$\ddot{\varphi} - \frac{g}{l}\sin(\varphi) = 0.$$

However, the problem of an inverted pendulum mounted on a cart with one degree of freedom is more complex. To derive the equations of motions for this setup the Lagrangian[162]

$$L = T - V \tag{6.85}$$

is used where T as usual represents the total kinetic energy while V is the potential energy of the system.[163] The total potential energy is

$$V = mgl\cos(\varphi), \tag{6.86}$$

with g representing the gravitational acceleration. The total kinetic energy is the sum of the kinetic energies of the pendulum bob having mass m and the moving cart with mass M, i.e.,

$$T = \frac{1}{2}\left(Mv_c^2 + mv_p^2\right) \tag{6.87}$$

---

[162] See [LEVI 2012, p. 115 et seq.].
[163] The following derivation basically follows https://en.wikipedia.org/wiki/Inverted_pendulum, retrieved March 29th, 2019.

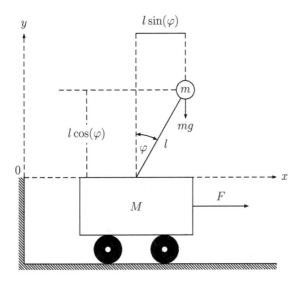

Fig. 6.116. Configuration of the inverted pendulum

where v_c is the velocity of the cart, and v_p is the velocity of the pendulum bob. Obviously,

$$v_c = \dot{x} \tag{6.88}$$

and (a bit less obviously)

$$v_p = \sqrt{\left(\frac{\mathrm{d}}{\mathrm{d}t}(x - l\sin(\varphi))\right)^2 + \left(\frac{\mathrm{d}}{\mathrm{d}t}(l\cos(\varphi))\right)^2}.$$

The left term under the square root is

$$(\dot{x} - l\dot{\varphi}\cos(\varphi))^2 = \dot{x}^2 - 2\dot{x}l\dot{\varphi}\cos(\varphi) + l^2\dot{\varphi}^2\cos^2(\varphi)$$

while the right term is

$$(-l\dot{\varphi}\sin(\varphi))^2 = l^2\dot{\varphi}^2\sin^2(\varphi)$$

yielding

$$\begin{aligned}v_p^2 &= \dot{x}^2 - 2\dot{x}l\dot{\varphi}\cos(\varphi) + l^2\dot{\varphi}^2\cos^2(\varphi) + l^2\dot{\varphi}^2\sin^2(\varphi)\\ &= \dot{x}^2 - 2\dot{x}\dot{\varphi}l\cos(\varphi) + l^2\dot{\varphi}^2\end{aligned} \tag{6.89}$$

since $\sin^2(\varphi) + \cos^2(\varphi) = 1$.

Substituting (6.86) and (6.87) into (6.85) using (6.88) and (6.89) yields the Lagrangian

$$L = \frac{1}{2}M\dot{x}^2 + \frac{1}{2}m\left(\dot{x}^2 - 2\dot{x}\dot{\varphi}l\cos(\varphi) + l^2\dot{\varphi}^2\right) - mgl\cos(\varphi)$$

$$= \frac{1}{2}(M+m)\dot{x}^2 - m\dot{x}\dot{\varphi}l\cos(\varphi) + \frac{1}{2}ml^2\dot{\varphi}^2 - mgl\cos(\varphi). \quad (6.90)$$

The EULER-LAGRANGE-equations of the general form

$$\frac{d}{dt}\left(\frac{\partial L}{\partial \dot{q}_i}\right) = \frac{\partial L}{\partial q_i}$$

yield

$$\frac{d}{dt}\left(\frac{\partial L}{\partial \dot{x}}\right) - \frac{\partial L}{\partial x} = F \text{ and}$$

$$\frac{d}{dt}\left(\frac{\partial L}{\partial \dot{\varphi}}\right) - \frac{\partial L}{\partial \varphi} = 0$$

with the generalized coordinates $q_1 = x$ and $q_2 = \varphi$. These in turn yield

$$\frac{\partial L}{\partial x} = 0,$$

$$\frac{\partial L}{\partial \dot{x}} = (M+m)\dot{x} - ml\dot{\varphi}\cos(\varphi) \text{ and thus}$$

$$\frac{d}{dt}\left(\frac{\partial L}{\partial \dot{x}}\right) = (M+m)\ddot{x} - ml\ddot{\varphi}\cos(\varphi) + ml\dot{\varphi}^2\sin(\varphi).$$

Similarly

$$\frac{\partial L}{\partial \varphi} = ml\dot{x}\dot{\varphi}\sin(\varphi) + mgl\sin(\varphi),$$

$$\frac{\partial L}{\partial \dot{\varphi}} = -ml\dot{x}\cos(\varphi) + ml^2\dot{\varphi} \text{ and}$$

$$\frac{d}{dt}\left(\frac{\partial L}{\partial \dot{\varphi}}\right) = -ml\ddot{x}\cos(\varphi) + ml\dot{x}\dot{\varphi}\sin(\varphi) + ml^2\ddot{\varphi}.$$

These finally result in

$$\frac{d}{dt}\left(\frac{\partial L}{\partial \dot{x}}\right) - \frac{\partial L}{\partial x} = (M+m)\ddot{x} - ml\ddot{\varphi}\cos(\varphi) + ml\dot{\varphi}^2\sin(\varphi) = F \quad (6.91)$$

and

$$\frac{d}{dt}\left(\frac{\partial L}{\partial \dot{\varphi}}\right) - \frac{\partial L}{\partial \varphi} = -ml\ddot{x}\cos(\varphi) + ml\dot{x}\dot{\varphi}\sin(\varphi) + ml^2\ddot{\varphi} - ml\dot{x}\dot{\varphi}\sin(\varphi) - mgl\sin(\varphi) = 0.$$

Dividing this last equation by ml yields

$$l\ddot{\varphi} - \ddot{x}\cos(\varphi) - g\sin(\varphi) = 0$$

which can be solved for $\ddot{\varphi}$ yielding

$$\ddot{\varphi} = \frac{1}{l}(\ddot{x}\cos(\varphi) + g\sin(\varphi)). \quad (6.92)$$

The two equations (6.91) and (6.92) fully describe the motion of an inverted pendulum mounted on a cart capable of moving along the x direction under an external force F and can now be used to derive an analog computer setup for this problem.

Assuming that the mass m of the pendulum bob is negligible compared to the mass M of the moving cart equation (6.91) can be simplified to

$$M\ddot{x} = F,$$

i.e., the movement of the pendulum mounted on the cart has no influence on the cart's movement. With this assumption the inverted pendulum problem can be described just by equation (6.92) assuming that

$$\ddot{x} = \frac{F}{M}.$$

Equation 6.92 can be directly converted into the corresponding analog computer program as shown in figure 6.117. Based on the input variable $\ddot{x}$ this setup generates the time $\dot{\varphi}$ derivative of the generalized coordinate φ, which is then used to generate the two required harmonic terms $\sin(\varphi)$ and $\cos(\varphi)$ without the need for a sine/cosine function generator, as shown in figure 6.118. Note that the term $\frac{1}{l}$ has been deliberately omitted by defining $l := 1$.[164] γ_1 controls how much the acceleration $\ddot{x}$ affects the pendulum while γ_2 controls its effect on the movement of the cart.[165]

The potentiometer labeled g yields the gravitational acceleration factor while β is an extension of the basic equation of motion and introduces a damping term for the angular velocity $\dot{\varphi}$ of the pendulum, making the simulation more realistic. Setting $\beta = 0$ results in a frictionless pendulum.

What would an analog computer setup be without a proper visualization of the dynamic system? The program shown in figure 6.119 displays the cart with its attached pendulum rod as it moves along the x-axis controlled by the input function $\ddot{x}$. To demonstrate the behavior of the system a double pole switch with neutral middle position can be used to generate a suitable input signal $\ddot{x}$ to push the cart to the left or right. With a suitable low time scale factor set on the integrators yielding $\cos(\varphi)$ and $\sin(\varphi)$ the pendulum can even be manually balanced for a short time, given some operator training.

To generate a flicker-free display an amplitude stabilized quadrature signal pair $\sin(\omega t)$ and $\cos(\omega t)$ of some kHz is required; this can be derived from the computer setup shown in figure 4.11 or 4.13 in section 4.2. This signal is used to draw a circular or elliptical cart as well as the rotating pendulum rod mounted on the cart.

---

[164] The output function $\ddot{\varphi}$ will be used later.
[165] These two parameters include the respective masses of the cart and the pendulum.

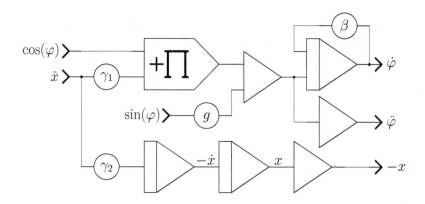

Fig. 6.117. Motion of the pendulum mounted on its cart

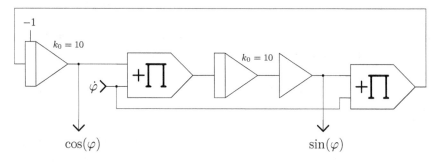

Fig. 6.118. Generating $\sin(\varphi)$ and $\cos(\varphi)$

The length of the rod is set by the potentiometer labeled l. Since the pendulum rod has one of its ends fixed at the center of the cart figure the $\sin(\omega t)$-signal has to be shifted by an amount r. The actual rotation is done by multiplication with $\sin(\varphi)$ and $\cos(\varphi)$. The circuit yields two output signal-pairs: P_x and P_y are used to draw the pendulum rod while C_x and C_y display the moving cart. Therefore, an oscilloscope with two independent (multiplexed) x, y-inputs is required for this simulation, if both, cart and pendulum, are to be displayed. A screenshot of a typical resulting display with the pendulum just tipping over is shown in figure 6.120.

What happens when the mass of the pendulum bob is no longer negligible? In this case equation (6.91) can no longer be simplified as before but must be fully taken into account to account for the influence of the swinging pendulum on the cart. Solving this equation for $\ddot{x}$ yields

$$\ddot{x} = \frac{1}{M+m}\left(F + ml\ddot{\varphi}\cos(\varphi) - ml\dot{\varphi}^2\sin(\varphi)\right)$$

which can be easily transformed into a computer setup shown in figure 6.121.

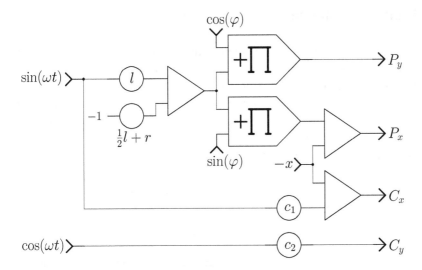

Fig. 6.119. Display of the pendulum mounted on its cart

Fig. 6.120. Display of the pendulum mounted on its cart

The function $\ddot{x}$ generated by this circuit is now fed into the associated input of the subcircuit shown in figure 6.117. An external force acting on the cart can be introduced by means of switch S1 while γ_3 controls the strength of this force.

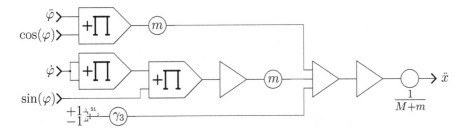

Fig. 6.121. Effect of a heavy mass pendulum on the moving cart

6.34 Elastic pendulum

The approach of using the Lagrangian in conjunction with the EULER-LAGRANGE-equations is also used here to model an elastic pendulum, i.e., a mass mounted on an elastic spring of stiffness k instead of a stiff pendulum rod. As before, φ describes the deflection of the pendulum from its position of rest. x_0 represents the length of the spring without any external forces being applied. Its elongation or contraction due to external forces is denoted by x.

The Lagrangian is $L = T - V$ with

$$V = V_{\text{spring}} + V_{\text{mass}} = \frac{1}{2}x^2 - gm(x_0 + x)\cos(\varphi)$$

and

$$T = \frac{1}{2}mv^2.$$

Since v consists of two components along the horizontal and vertical axes of the coordinate system we get

$$T = \frac{1}{2}m\dot{x}^2 + \frac{1}{2}(x_0 + x)^2 \dot{\varphi}^2 = \frac{1}{2}m\left(\dot{x}^2 + (x_0 + x)^2 \dot{\varphi}^2\right).$$

This yields the Lagrangian

$$L = \frac{1}{2}m\dot{x}^2 + \frac{1}{2}m(x_0 + x)^2 \dot{\varphi}^2 - \frac{1}{2}kx^2 + gm(x_0 + x)\cos(\varphi).$$

The EULER-LAGRANGE-equations

$$\frac{d}{dt}\left(\frac{\partial L}{\partial \dot{q}_i}\right) - \frac{\partial L}{\partial q_i} = 0$$

with the generalized coordinates $q_1 = x$ and $q_2 = \varphi$ yield

$$\frac{d}{dt}\left(\frac{\partial L}{\partial \dot{x}}\right) - \frac{\partial L}{\partial x} = -m\ddot{x} - m(x_0 + x)\dot{\varphi}^2 + kx - gm\cos(\varphi) \text{ and}$$

$$\frac{d}{dt}\left(\frac{\partial L}{\partial \dot{\varphi}}\right) - \frac{\partial L}{\partial \varphi} = 2m(x_0 + x)\dot{x}\dot{\varphi} + m(x_0 + x)^2 \text{varphi} - gm(x_0 + x)\sin(\varphi).$$

Solving for $\ddot{x}$ and $\ddot{\varphi}$ results in

$$\ddot{x} = (x_0 + x)\dot{\varphi}^2 - \frac{k}{m}x + g\cos(\varphi) \text{ and} \tag{6.93}$$

$$\ddot{\varphi} = -\frac{2}{x_0 + x}\dot{x}\dot{\varphi} - \frac{g}{x_0 + x}\sin(\varphi). \tag{6.94}$$

To simplify things a bit $m = 1$ is assumed, which reduces the term $\frac{k}{m}$ to k. Setting $l = x_0 + x$ simplifies equation (6.93) to

$$\ddot{x} = l\dot{\varphi}^2 - kx + g\cos(\varphi).$$

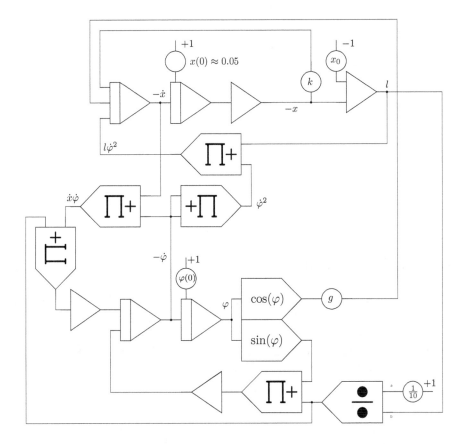

Fig. 6.122. Elastic pendulum simulation

Scaling the term $2/l$ in equation (6.94) with $\lambda = \frac{1}{20}$ and setting $g = \frac{1}{10}$ yields two identical fractions in equation (6.94) yielding

$$\ddot{\varphi} = -\lambda \frac{2}{l} \dot{x}\dot{\varphi}\frac{1}{\lambda} - \frac{g}{l}\sin(\varphi) = -\frac{1}{10l}\left(\dot{x}\dot{\varphi}\frac{1}{\lambda} + \sin(\varphi)\right).$$

Figure 6.122 shows the corresponding analog computer program. Note that only one divider is required due to the trick of setting $g = \frac{1}{10}$. This not only saves a computing element but also reduces errors.

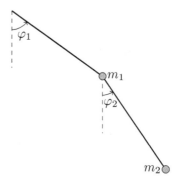

Fig. 6.123. Double pendulum

6.35 Double pendulum

Even more complicated, but also more mesmerizing, is the simulation of the double pendulum shown in figure 6.123 on an analog computer.[166]

The equations of motion will again be derived by determining the Lagrangian $L = T - V$ with the two generalized coordinates φ_1 and φ_2. T represents the total kinetic energy while V is the potential energy of the system. The potential energy is

$$V = -g\left((m_1 + m_2)l_1 \cos(\varphi_1) + m_2 l_2 \cos(\varphi_2)\right) \tag{6.95}$$

with m_1 and m_2 representing the masses of the two bobs mounted on the tips of the two pendulum rods (which are assumed to be weightless). As always, g represents the gravitational acceleration.

To derive the total kinetic energy the positions of the pendulum arm tips (x_1, y_1) and (x_2, y_2) are required:

$$x_1 = l_1 \sin(\varphi_1)$$
$$y_1 = l_1 \cos(\varphi_1)$$
$$x_2 = l_1 \sin(\varphi_1) + l_2 \sin(\varphi_2)$$
$$y_2 = l_1 \cos(\varphi_1) + l_2 \cos(\varphi_2)$$

The first derivatives of these with respect to time are

$$\dot{x}_1 = \dot{\varphi}_1 l_1 \cos(\varphi_1),$$
$$\dot{y}_1 = -\dot{\varphi}_1 l_1 \sin(\varphi_1),$$

---

[166] This has also been done by [MAHRENHOLTZ 1968, pp. 159–165] which served as an inspiration for this example, especially with respect to the display circuit.

$$\dot{x}_2 = \dot{\varphi}_1 l_1 \cos(\varphi_1) + \dot{\varphi}_2 l_2 \cos(\varphi_2), \text{ and}$$
$$\dot{y}_2 = -\dot{\varphi}_1 l_1 \sin(\varphi_1) - \dot{\varphi}_2 l_2 \sin \varphi_2.$$

The kinetic energy of the system is then

$$T = \frac{1}{2}\left(m_1(\dot{x}_1^2 + \dot{y}_1^2) + m_s(\dot{x}_2^2 + \dot{y}_2^2)\right) \tag{6.96}$$

with

$$\dot{x}_1^2 = \dot{\varphi}_1^2 l_1^2 \cos^2(\varphi_1),$$
$$\dot{y}_1^2 = \dot{\varphi}_1^2 l_1^2 \sin^2(\varphi_1),$$
$$\dot{x}_2^2 = \dot{\varphi}_1^2 l_1^2 \cos^2(\varphi_1) + 2\dot{\varphi}_1 \dot{\varphi}_2 l_1 l_2 \cos(\varphi_1)\cos(\varphi_2) + \dot{\varphi}_2^2 l_2^2 \cos^2(\varphi_2),$$
$$\dot{y}_2^2 = \dot{\varphi}_1^2 l_1^2 \sin^2(\varphi_1) + 2\dot{\varphi}_1 \dot{\varphi}_2 l_1 l_2 \sin(\varphi_1)\sin(\varphi_2) + \dot{\varphi}_2^2 l_2^2 \sin^2(\varphi_2),$$

and thus

$$\dot{x}_1^2 + \dot{y}_1^2 = \dot{\varphi}_1^2 l_1^2 \cos^2(\varphi_1) + \dot{\varphi}_1^2 l_1^2 \sin^2(\varphi_1)$$
$$= \dot{\varphi}_1^2 l_1^2 \left(\cos^2(\varphi_1) + \sin^2(\varphi_1)\right)$$
$$= \dot{\varphi}_1^2 l_1^2, \text{ and} \tag{6.97}$$
$$\dot{x}_2^2 + \dot{y}_2^2 = \dot{\varphi}_1^2 l_1^2 \left(\cos^2(\varphi_1) + \sin^2(\varphi_1)\right) + \dot{\varphi}_2^2 l_2^2 \left(\cos^2(\varphi_2) + \sin^2(\varphi_2)\right) +$$
$$2\dot{\varphi}_1 \dot{\varphi}_2 l_1 l_2 \left(\cos(\varphi_1)\cos(\varphi_2) + \sin(\varphi_1)\sin(\varphi_2)\right)$$
$$= \dot{\varphi}_1^2 l_1^2 + \dot{\varphi}_2^2 l_2^2 + 2\dot{\varphi}_1 \dot{\varphi}_2 l_1 l_2 \cos(\varphi_1 - \varphi_2). \tag{6.98}$$

The Lagrangian L results from equations (6.95) and (6.96) with (6.97) and (6.98) as

$$L = \frac{1}{2}l_1^2 \dot{\varphi}_1^2(m_1 + m_2) + \frac{1}{2}m_2 \dot{\varphi}_2^2 l_2^2 + m_2 \dot{\varphi}_1 \dot{\varphi}_2 l_1 l_2 \cos(\varphi_1 - \varphi_2) +$$
$$g(m_1 + m_2)l_1 \cos(\varphi_1) + g m_2 l_2 \cos(\varphi_2). \tag{6.99}$$

Now the EULER-LAGRANGE-equations

$$\frac{d}{dt}\frac{\partial L}{\partial \dot{\varphi}_1} - \frac{\partial L}{\partial \varphi_1} = 0 \text{ and} \tag{6.100}$$

$$\frac{d}{dt}\frac{\partial L}{\partial \dot{\varphi}_2} - \frac{\partial L}{\partial \varphi_2} = 0 \tag{6.101}$$

have to be solved. The required (partial) derivatives are

$$\frac{\partial L}{\partial \varphi_1} = -m_2 \dot{\varphi}_1 \dot{\varphi}_2 l_1 l_2 \sin(\varphi_1 - \varphi_2) - g(m_1 + m_2)l_1 \sin(\varphi_1),$$

$$\frac{\partial L}{\partial \dot{\varphi}_1} = l_1^2 \dot{\varphi}_1(m_1 + m_2) + m_2 \dot{\varphi}_2 l_1 l_2 \cos(\varphi_1 - \varphi_2),$$

$$\frac{d}{dt}\frac{\partial L}{\partial \dot{\varphi}_1} = l_1^2 \ddot{\varphi}_1(m_1 + m_2) + m_2 l_1 l_2 \left[\ddot{\varphi}_2 \cos(\varphi_1 - \varphi_2) - \dot{\varphi}_2 \sin(\varphi_1 - \varphi_2)(\dot{\varphi}_1 - \dot{\varphi}_2)\right],$$

$$\frac{\partial L}{\partial \varphi_2} = m_2 \dot{\varphi}_1 \dot{\varphi}_2 l_1 l_2 \sin(\varphi_1 - \varphi_2) - g m_2 l_2 \sin(\varphi_2),$$

$$\frac{\partial L}{\partial \dot{\varphi}_2} = m_2 \dot{\varphi}_2 l_2^2 + m_2 \dot{\varphi}_1 l_1 l_2 \cos(\varphi_1 - \varphi_2), \text{ and}$$

$$\frac{d}{dt}\frac{\partial L}{\partial \dot{\varphi}_2} = m_2 \ddot{\varphi}_2 l_2^2 + m_2 l_1 l_2 [\ddot{\varphi}_1 \cos(\varphi_1 - \varphi_2) - \dot{\varphi}_1 \sin(\varphi_1 - \varphi_2)(\dot{\varphi}_1 - \dot{\varphi}_2)]$$

based on (6.99). Substituting these into (6.100) yields

$$0 = l_1^2 \ddot{\varphi}_1 (m_1 + m_2) - m_2 l_1 l_2 \ddot{\varphi}_2 \cos(\varphi_1 - \varphi_2) - m_2 l_1 l_2 \dot{\varphi}_2 \sin(\varphi_1 - \varphi_2)(\dot{\varphi}_1 - \dot{\varphi}_2) + \\ m_2 \dot{\varphi}_1 \dot{\varphi}_2 l_1 l_2 \sin(\varphi_1 - \varphi_2) + g l_1 (m_1 + m_2) \sin(\varphi_1).$$

Expanding $(\dot{\varphi}_1 - \dot{\varphi}_2)$ yields

$$0 = l_1^2 \ddot{\varphi}_1 (m_1 + m_2) + m_2 l_1 l_2 \ddot{\varphi}_2 \cos(\varphi_1 - \varphi_2) + \\ m_2 l_1 l_2 \dot{\varphi}_2^2 \sin(\varphi_1 - \varphi_2) + g l_1 (m_1 + m_2) \sin(\varphi_1).$$

Dividing by $l_1^2 (m_1 + m_2)$ results in

$$0 = \ddot{\varphi}_1 + \frac{m_2}{m_1 + m_2} \frac{l_2}{l_1} \ddot{\varphi}_2 \cos(\varphi_1 - \varphi_2) + \frac{m_2}{m_1 + m_2} \frac{l_2}{l_1} \dot{\varphi}_2^2 \sin(\varphi_1 - \varphi_2) + \frac{g}{l_1} \sin(\varphi_1),$$

which can be further simplified as

$$0 = \ddot{\varphi}_1 + \frac{m_2}{m_1 + m_2} \frac{l_2}{l_1} \left[\ddot{\varphi}_2 \cos(\varphi_1 - \varphi_2) + \dot{\varphi}_2^2 \sin(\varphi_1 - \varphi_2) \right] + \frac{g}{l_1} \sin(\varphi_1). \quad (6.102)$$

Proceeding analogously, (6.101) eventually yields

$$0 = \ddot{\varphi}_2 + \frac{l_1}{l_2} \left[\ddot{\varphi}_1 \cos(\varphi_1 - \varphi_2) - \dot{\varphi}_1^2 \sin(\varphi_1 - \varphi_2) \right] + \frac{g}{l_2} \sin(\varphi_2). \quad (6.103)$$

To simplify things further, assuming that $m_1 = m_2 = 1$ and $l_1 = l_2 = 1$ in arbitrary units yields the following two equations of motion for the double pendulum based on (6.102) and (6.103):

$$\ddot{\varphi}_1 = -\frac{1}{2} \left[\ddot{\varphi}_2 \cos(\varphi_1 - \varphi_2) + \dot{\varphi}_2^2 \sin(\varphi_1 - \varphi_2) + 2g \sin(\varphi_1) \right] \text{ and} \quad (6.104)$$

$$\ddot{\varphi}_2 = -\left[\ddot{\varphi}_1 \cos(\varphi_1 - \varphi_2) - \dot{\varphi}_1^2 \sin(\varphi_1 - \varphi_2) + g \sin(\varphi_2) \right]. \quad (6.105)$$

The implementation of equations (6.104) and (6.105) is straightforward, as shown in figures 6.124 and 6.125. Both circuits require the functions $\sin(\varphi_1 - \varphi_2)$, $\cos(\varphi_1 - \varphi_2)$, $\sin(\varphi_1)$, and $\sin(\varphi_2)$, which are generated by three distinct subcircuits like the one shown in figure 6.126. These circuits yield sine and cosine of an angle based on the first time-derivative of this angle, so the simulation is not limited to a finite range for the two angles φ_1 and φ_2.

It should be noted that each of the two integrators of each of these three harmonic function generators has a potentiometer connected to its respective initial value input. When a simulation run with given initial values for $\varphi_1(0)$ and

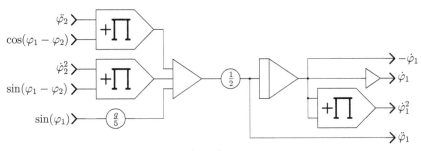

Fig. 6.124. Implementation of equation (6.104)

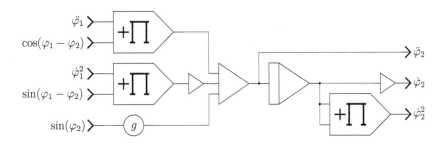

Fig. 6.125. Implementation of equation (6.105)

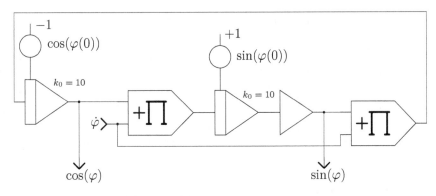

Fig. 6.126. Generating $\sin(\varphi)$ and $\cos(\varphi)$

$\varphi_2(0)$ is started, these potentiometers have to be set to $\cos(\varphi_1(0))$, $\sin(\varphi_1(0))$ and $\cos(\varphi_2(0))$, $\sin(\varphi_2(0))$, respectively.

Since $\dot\varphi_1$ and $\dot\varphi_2$ are readily available from the circuits shown in figures 6.124 and 6.125, two of these harmonic function generators can be directly fed with these values. The input for the third function generator is generated by a two-input summer as shown in figure 6.127.

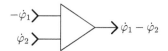

Fig. 6.127. Computing $\dot\varphi_1 - \dot\varphi_2$

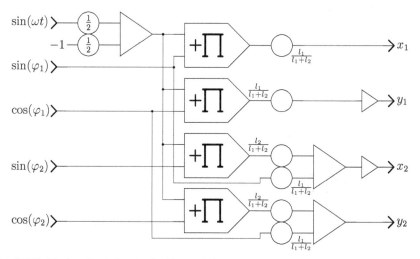

Fig. 6.128. Display circuit for the double pendulum

With $\dot\varphi_1$ and $\dot\varphi_2$ and thus $\sin(\varphi_1)$, $\cos(\varphi_1)$, and $\sin(\varphi_2)$, $\cos(\varphi_2)$ readily available the double pendulum can be displayed on an oscilloscope featuring two separate x, y-inputs by means of the circuit shown in figure 6.128. This requires, as before, a high-frequency input $\sin(\omega t)$, which can be generated as usual.

Figure 6.129 shows a long-term exposure of the movements of the double pendulum starting as an inverted pendulum with its first pendulum rod pointing upwards while the second one points downwards. This simulation requires ten integrators, 15 summers, 16 multipliers, and 17 coefficient potentiometers.

6.36 Making Music

Being remarkably similar to a music synthesizer the idea of making music with an analog computer is quite obvious. This section describes a simple computer

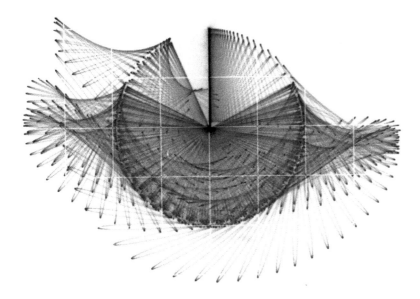

Fig. 6.129. Long-term exposure of a double pendulum simulation run

setup that turns an analog computer into a monophonic synthesizer controlled by a keyboard generating voltage outputs for gate and pitch.[167]

Figure 6.130 shows the overall setup for an analog computer to be used as a simple monophonic synthesizer. The circuit on top is a simple triangular/square wave generator. The "mod."-input is the modulation voltage which is expected to be in the interval $[0, 1]$. Many synthesizers use a saw tooth waveform as the basis for sound creation due to the high harmonic content of such a signal which can be filtered out and mixed in a suitable fashion to yield all kinds of fascinating sounds. The triangular output signal generated here deviates from this quite a bit but is sufficient for a little demonstration.

The circuit in the middle of the figure has a "gate" input which switches the output on and off if a key is depressed and released. The three position switch is used to select which waveform is fed to the audio amplifier connected to the output labeled "out". In the upper and lower position the switch just selects the triangular or square wave output from the oscillator circuit to the output. In its second position from top it connects the output of a variable function generator $f(x)$ to

---

[167] In August 2020, HAINBACH, HANS KULK, and BERND ULMANN streamed a live discussion titled "Analog computers for music", which can be found here: https://www.youtube.com/watch?v=bgyzeyatS-0.

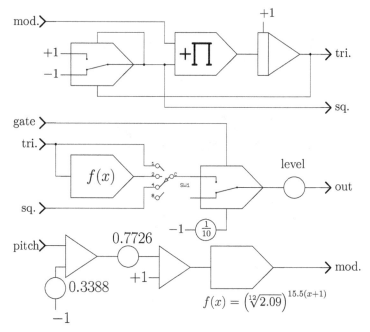

Fig. 6.130. Analog computer setup as monophonic synthesizer

the output. This function generator is connected to the triangular wave output and can be used to generate fairly arbitrary wave forms which yield interesting sounds. The fact that it is fed a triangular wave instead of a saw tooth signal ensures that the function set on the function generator is played in a symmetric way, thereby guaranteeing that no steps resulting in audible clicks occur between the end of one period of the output waveform and the start of the next.

The most interesting subcircuit is shown in the lower third of figure 6.130. The pitch output of a typical keyboard varies linearly with 1 V/octave which is convenient from a keyboard point of view but does not match the requirement of a linearly increasing control voltage for the oscillator as that described above.

The reason for this logarithmic voltage/frequency relationship is due to our Western musical culture where the frequency of a tone doubles from one octave to the next. Given our traditional half-tone scheme with twelve half-tones comprising one octave, the increase in frequency from one half-tone to the next is $\sqrt[12]{2}$. Accordingly, the voltage output of the keyboard must be fed into an exponential function generator yielding a suitable control voltage for the oscillator. Here things get a bit involved: The keyboard used in this example is pretty small with 32 half-tones and its linear output voltage, which is always positive, must be mapped to the analog computer value interval of $[-1, 1]$ in order to make best use of the variable function generator used to implement the exponential function.

x	$g(x)$	x	$g(x)$
-1.0	0.0700	1.0	0.4700
-0.9	0.0769	0.9	0.4274
-0.8	0.0846	0.8	0.3885
-0.7	0.0931	0.7	0.3533
-0.6	0.1024	0.6	0.3212
-0.5	0.1126	0.5	0.2920
-0.4	0.1239	0.4	0.2654
-0.3	0.1363	0.3	0.2414
-0.2	0.1499	0.2	0.2195
-0.1	0.1649	0.1	0.1995
0	0.1814		

Table 6.3. Function generator setting for equation (6.106)

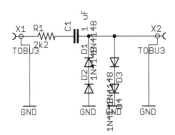

Fig. 6.131. Simple audio adapter circuit

This mapping is done by the two summers shown in the lower sub-circuit of figure 6.130. The function generator actually implements

$$g(x) = \frac{7}{100} \left(\sqrt[12]{2.09} \right)^{15.5(x+1)} \tag{6.106}$$

instead of using $\sqrt[12]{2}$ as the basis. This is due to the fact that the implementation of the triangular wave oscillator suffers a tiny bit from the unavoidable hysteresis of the electronic comparator and its associated switch. Table 6.3 shows the 21 interpolation points of the function generator used to implement $g(x)$.

Figure 6.131 shows the (very simple) audio adapter connecting the output of the analog computer to a conventional amplifier or a computer sound card. The two strings of 1N4148 diodes limit the signal amplitude to about 1.4 V to avoid damage to the amplifier or sound card in case of excessive levels. Figure 6.132 gives an impression of the overall setup.

Fig. 6.132. Impression of the actual setup on a Telefunken RA 770

6.37 Neutron kinetics

This section shows a simple implementation of neutron kinetics for the simulation of nuclear reactors fuelled with ^{235}U or ^{239}Pu respectively.[168] Following [FENECH et al. 1973, p. 128] neutron kinetics in such a reactor can be described by

$$\dot{n} = \frac{n}{l^*}(\delta K - \beta) + \sum_i \lambda_i c_i + s \text{ and} \qquad (6.107)$$

$$\dot{c}_i = \frac{n\beta_i}{l^*} - \lambda_i c_i$$

with n being the neutron density, s representing an external neutron source (which is often used to "start" a nuclear reactor), and l^* representing the effective neutron lifetime, which depends on the reactor design and geometry and can range between values as big as 10^{-2} s and 10^{-4} s.[169] c_i are the precursors of the i-th group of delayed neutrons, β_i represents the fraction of the i-th delayed neutron group, and λ_i is the decay constant for the i-th precursor group.

Typically, six groups of delayed neutron precursors are taken into account for a nuclear reactor simulation. Equation (6.107) could be programmed in a straightforward way using one integrator for each neutron group. This section shows how a specialized computing element can be devised from actual data on

---

[168] An overview on nuclear reactor simulation techniques can be found in [MORRISON 1962]. Details of neutron kinetics and related topics are described in depth in [WEINBERG et al. 1958]. An actual implementation of a nuclear power plant simulator is shown in [FENECH et al. 1973].
[169] s will be ignored in the following – an external neutron source can always be modelled by adding s to the input of the network described here.

		^{235}U				^{239}Pu		
i	λ_i	β_i	$C_i = \frac{\eta\beta_i}{\lambda_i}$	R_i [Ω]	λ_i	β_i	$C_i = \frac{\eta\beta_i}{\lambda_i}$	R_i [Ω]
1	0.0127	0.000237	1.9 μF	414000	0.0129	0.0000798	619 nF	1252332
2	0.0317	0.001385	4.37 μF	72187	0.0311	0.000588	1.89 μF	170128
3	0.115	0.001222	1.06 μF	82034	0.134	0.0004536	339 nF	220138
4	0.311	0.002645	0.85 μF	37828	0.331	0.0006881	208 nF	145248
5	1.4	0.000832	59 nF	121065	1.26	0.0002163	17 nF	466853
6	3.87	0.000169	4.4 nF	587268	3.21	0.0000734	2.2 nF	1416029

Table 6.4. Delayed neutron data for ^{235}U and ^{239}Pu reactor

nuclear processes that implements the neutron kinetics of a nuclear reactor with reasonable accuracy and saves a lot of common computing elements.

Figure 6.133 shows the basic structure of a such a computing element implementing the neutron kinetics for either a ^{235}U or ^{239}Pu reactor. The basic circuit consisting of RL and either CL1 or CL2 is an integrator with

$$l^* = \text{RL} \cdot \text{CL}_j, \ 1 \leq j \leq 2.$$

Choosing RL= 100 kΩ and CL1= 1 nF yields $l^* = 10^{-4}$s. With the same RL a CL2 = 100 nF results in $l^* = 10^{-2}$s, two suitable and not too unrealistic values for the effective neutron lifetime in a nuclear reactor. The switch S1 selects which of these two values is to be used in a simulation run.

The six groups of delayed neutron precursors and resulting neutrons can be modelled by series circuits consisting of RU$_i$ and CU$_i$ for the case of ^{235}U and RP$_i$ and CP$_i$ for ^{239}Pu respectively. The capacitor values are determined by[170]

$$C^*_i = \eta \frac{\beta_i}{\lambda_i}$$

with η being a scaling factor to get reasonable values for the capacitors. In this case $\eta = 10^2$ was chosen. With the capacitances determined this way, the resistors can be determined according to

$$R^*_i = \frac{1}{\eta \lambda_i C^*_i}.$$

Table 6.4 shows the λ_i and β_i for ^{235}U and ^{239}Pu with the corresponding values for the resistor-capacitor networks for both cases.[171]

---

[170] Cf. [FENECH et al. 1973, p.132]. C^*_i and R^*_i denote the capacitor and resistor values for either the ^{235}U or ^{239}Pu case.

[171] The values λ_i and β_i are according to [TYROR et al. 1970, p. 22]. Actual values tend to differ a bit in the literature.

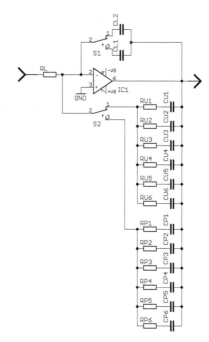

Fig. 6.133. Basic circuit for the simulation of neutron kinetics with six groups of delayed neutrons

The circuit shown in figure 6.133 can be used directly in an analog computer setup although it is advisable to add two electronic switches, one of which connects RL to the summing junction of the operational amplifier only when the computer is in operate mode, while the other switch discharges the selected capacitor CL1 or CL2 as well as the selected network for the delayed neutron groups when the computer is in initial condition mode. Due to the relatively large values of the series connection of resistors and capacitors, the initial condition time should be chosen to be long enough to ensure that all capacitors are suitably discharged.

6.38 Smooth sorting

The final example in this chapter is particularly interesting as it defies intuition by solving an inherently discrete problem, namely sorting values on an analog computer – a process also known as *smooth sorting*.[172] It is based on the following set of coupled differential equations:

---

[172] This section is based on [BLOCH et al. 2010], [ZHAN et al. 2016], http://www.hrl.harvard.edu/analog/, retrieved December 12th, 2022, and [BROCKETT 1991].

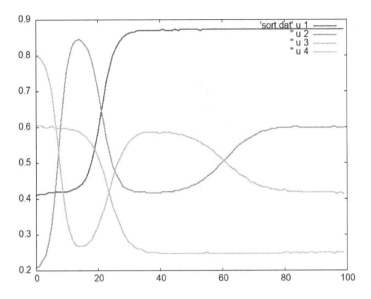

Fig. 6.134. Behavior of the analog sorting program

$$\dot{x}_1 = 2y_1^2$$
$$\dot{y}_1 = -(x_1 - x_2)y_1$$
$$\dot{x}_2 = 2(y_2^2 - y_1^2)$$
$$\dot{y}_2 = -(x_2 - x_3)y_2$$
$$\ldots$$
$$\dot{x}_n = -2y_{n-1}^2$$

These directly yield the program for sorting four values shown in figure 6.135. Figure 6.134 shows the result of a typical program run.

The initial conditions $y_i(0)$ should be "small", about $\frac{1}{100}$ has proven to work well. The values to be sorted are given by $x_i(0)$. Figure 6.134 shows a typical result from such a smooth sorting run. It should be noted that this approach is very sensitive even to tiny offsets introduced by the multipliers, which will cause the results to drift away quickly.

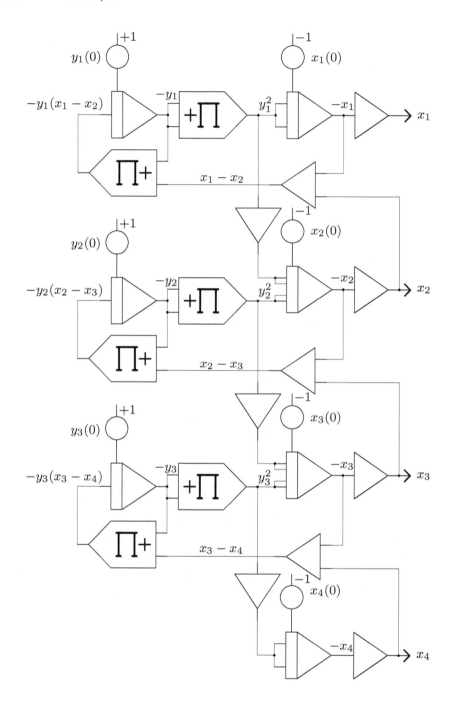

Fig. 6.135. Smooth sorting program

7
Hybrid computing

Although analog computers are extremely powerful on their own, the combination of an analog computer and a digital computer forming a *hybrid computer* is even more powerful.

This idea is by no means new – as early as 1956 *Space Technology Laboratories* developed the *ADDAVERTER*, a system capable of coupling an analog computer with a digital computer by using up to 15 ADCs and DACs, each featuring a resolution of ± 0.1 %. The rationale behind such a setup is that digital computers are good in areas where analog computer have deficiencies and vice versa. Applications for such hybrid computer setups are abundant due to their computational power and energy efficiency.

A hybrid computer typically operates in one of two basic modes:

Alternating operation: In this mode of operation the digital computer typically controls the parameters of an analog computer setup. The analog computer performs a run based on a particular parameter set and returns the results of its computation to the digital computer by means of ADCs. Based on these results the digital computer can then derive a new set of parameters and so forth. Since the operation of the digital and analog computers do not overlap, this mode is not time critical with respect to communication latencies, ADCs conversion times, etc. Alternating operation is typically employed for the solution of optimization problems, multidimensional parameter searches, the solution of partial differential equations, stochastic analyses, and related problems.

Parallel operation: This mode of operation requires real time operation from the digital computer as it operates in parallel with the analog computer and has to read and supply values with as little latency as possible. In a setup like this, the digital computer can not only change coefficients at run-time but also generate functions,[173] act as a delay circuit, make decisions based on the values of analog variables, and much more.

7.1 Hybrid controllers

The following examples and explanations are based on the hybrid controller for the Analog Paradigm Model-1 analog computer, but the ideas and techniques presented are by not restricted to this particular setup.[174] Similar coupling devices have been available in the past and can be built using cheap off-the-shelf hardware such as Arduino®-boards, etc.[175] A hybrid controller like this will typically provide the following functions:

Control lines: The hybrid controller must be able to control the mode of operation of the analog computer's integrators. In the simplest case it will accept commands like ic, op, or halt from the digital computer and set the ModeIC and ModeOP control lines accordingly.

Depending on its connection to the digital computer there might be considerable delays in the exchange of signals due to communication latencies.[176] This can be problematic in applications where precise timing of IC- and OP-times is required. It is therefore recommended that the hybrid controller itself should feature a clock with at least a 1 μs resolution. All timing issues should be performed locally by the hybrid controller without the need to communicate with the attached digital computer.

The hybrid controller should also be able to sense overload conditions, to control the POTSET line,[177] and it should feature an input line for an external halt signal.

Digital potentiometers: This type of potentiometer, similar to a *multiplying DAC*, typically offers 10 bit resolution, allowing parameters to be set with a resolution of about 0.1%.

If the input of such a potentiometer is connected to +1 or −1, it can even be used as a simple and cheap DAC allowing the digital computer not only

---

[173] Using a digital computer to generate functions of several variables is very useful as this is notoriously difficult to implement on a pure analog computer.
[174] Appendix C describes a simple hybrid controller for THE ANALOG THING.
[175] The schematics of a simple Arduino® based hybrid controller can be found at http://analogmuseum.org/english/examples/hybrid/, retrieved on March 10th, 2020.
[176] Latencies of up to several tens of ms are not unusual for serial communication over USB depending on the operating system, device drivers, etc.
[177] See section 2.5.

to control parameters but also to feed time-varying values to the analog computer.

Digital inputs: Many applications, such as parameter optimization, include some conditions which have to be sensed by the digital computer in order to abort a computer run or to change some parameters. This is typically achieved by means of comparators yielding a digital output signal which can be read by the digital computer using a digital input channel of the hybrid controller.

Digital outputs: Using digital output lines of the hybrid controller, electronic switches, such as those used in the CMP4 module, can be controlled by the attached digital computer. A typical application for this is to apply step functions to a simulated system.

Readout: One of the most important functions that a hybrid controller must implement is a readout capability including provisions for data logging. Ideally, the analog computer features a central readout system which allows every computing element to be addressed unambiguously. The hybrid controller can then address an individual computing element which connects its output to a central ADC for readout.

The time required for a single readout operation is highly critical and includes the time to address and select the computing element, the time required by the ADC for one conversion, and the time for transmitting the digital value to the digital computer.

If many elements have to be read during a computation over and over again, it can be necessary to place the analog computer in HALT-mode before starting readout operations and to reactivate OP-mode afterwards in order to avoid excessive skew between the individual data values being read. This becomes more and more important with increasing k_0 values (time scaling) of the integrators in a given setup.

A more sophisticated hybrid controller may also offer a number of ADC channels possibly fitted with sample-hold inputs, which can be connected to the outputs of computing elements. This allows all channels to be sampled at once thus eliminating timing skew between the various channels.

Figure 7.1 shows the front panel of the Analog Paradigm hybrid controller (HC). The group of 16 2 mm jacks in the upper half is connected to eight digital potentiometers each with a resolution of 10 bits. The second group of 16 jacks in the lower half connects to eight digital input lines and eight digital output lines. On the left is the USB port used to connect to the digital host computer, while an additional HALT-input and a trigger output are available on the lower right.

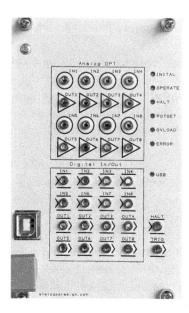

Fig. 7.1. Hybrid controller

The module controls all address, data, and control lines of the analog computer's system bus[178] and features a 16 bit ADC for readout.

The following examples give a basic overview of the general capabilities of such a hybrid computer. More information on hybrid computing in general with many applications can be found in [BEKEY et al. 1968], [KORN et al. 1964], [SCHÖNEFELD 1977], and other classic texts.

7.2 Basic operation

The Model-1's hybrid controller used in these examples is connected to the digital computer by means of a USB interface which emulates a serial line.[179] To facilitate programming the digital computer a Perl[180]-module (IO::HyCon) has been implemented; it provides an object oriented interface to the hybrid controller.[181]

---

[178] See section H.
[179] It is therefore possible to manually control the analog computer equipped with this controller instead of a manual control unit by sending appropriate commands and parameters using a serial line terminal application. See appendix I for a description of these commands.
[180] See [CHROMATIC 2015] for an introduction to modern Perl programming.
[181] A description of this module can be found in https://metacpan.org/pod/IO::HyCon, retrieved April 4th, 2020. As of 2023 there now also exists a Python package for the same purpose: https://github.com/anabrid/pyanalog, retrieved February 1st, 2023.

IO::HyCon requires a configuration file in YAML-format[182] describing the actual setup of the analog computer, i. e., the addresses of the various computing elements, the serial line parameters, etc. A typical minimum example is shown in the following listing:[183]

```
                                                example.yml
serial:
    port: /dev/cu.usbserial-DN05OL1O
    bits: 8
    baud: 115200
    parity: none
    stopbits: 1
    poll_interval: 10
    poll_attempts: 20000
types:
    0: PS
    1: SUM8
    2: INT4
    3: PT8
    4: CU
    5: MLT8
    6: MDS2
    7: CMP4
    8: HC
manual_potentiometers:
elements:
    INT-0: 0x0030
    INT-1: 0x0031
    INT-2: 0x0032
    INT-3: 0x0033
                                                example.yml
```

It consists of the following sections:

serial: This section defines the communication setup. port defines the serial device. Its name depends on the operating system and the USB-serial-adapter being used. The remaining parameters are defaults and can be adopted unchanged in most cases.

---

182 See https://yaml.org/spec/1.2/spec.html (retrieved October 3rd, 2019) for a detailed description of YAML.
183 There are more optional sections and settings which will be introduced as required.

types: Each computing element in an Analog Paradigm Model-1 returns an eight bit ID on the data bus when selected for readout. This section contains the mapping between these IDs and the actual computing element types.

manual_potentiometers: This section allows the definition of manual potentiometers which can be read by the hybrid controller. This is useful if a simulation requires user input by means of coefficient potentiometers whose values are of interest during or after an analog computer run. In most cases this section can be left empty.

elements: All elements which are to be read out by the hybrid controller should be defined in this section. Names like INT-0 should be avoided in favour of meaningful variable names such as phi-dot or the like.

A typical hybrid computer program now begins like this:

```
                                                skeleton.pl
1  use strict;
2  use warnings;
3  use IO::HyCon;
4
5  my $ac = IO::HyCon->new();
                                                skeleton.pl
```

The scalar variable $ac is an object with access to all the methods implemented in IO::HyCon to control the operation and setup of the analog computer. A call $ac->ic() will set the analog computer to IC-mode, $ac->op() will initiate an OP phase, and so on.

7.3 Shell trajectory

This first example is based on the ballistic trajectory simulation described in section 6.9 and is an example of digital and analog computers working in an alternating fashion. The analog computer setup shown in figure 6.31 is used in this hybrid computer setup. The only difference here is that two of the hybrid controller's eight digital potentiometers are used for the coefficients $\dot{x}(0) = \cos(\varphi)$ and $\dot{y}(0) = \sin(\varphi)$ with $\dot{x}(0)$ corresponding to DPT0 and $\dot{y}(0)$ to DPT1.

The aim of this simulation is to determine the required elevation angle of the cannon in this two-dimensional setup so that a target at a user defined x-coordinate will be hit. The digital computer sets an elevation angle φ by means of $\dot{x}(0)$ and $\dot{y}(0)$, starts a simulation run, and reads the x component where the shell hit the ground. Based on $\delta = x - x_{\text{target}}$, the distance between the target position and this point a new φ is determined and so on until δ is sufficiently small.

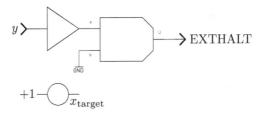

Fig. 7.2. HALT detection and target input

It is assumed that the y component of the cannon's position satisfies $y_{cannon} > 0$ while the target's y component is $y_{target} = 0$. Using a comparator as shown in figure 7.2, an external halt signal is generated when the shell hits the ground. This signal is connected to the EXTHALT input of the hybrid controller. x_{target} can be set manually by the potentiometer shown below the comparator. Its output is not connected to anything as it is only read out by the digital computer after a run to determine the miss distance δ.

The configuration file for this hybrid simulation is shown below – the sections serial and types have been omitted as these are identical to those in the listing shown on page 223. The two computing elements yielding the values required by the digital portion of the hybrid computer setup are defined in the elements section and labeled X and TARGET, representing x_{shell} and x_{target}.

──────────── trajectory.pl ────────────
```
elements:
    X:      0x0221
    TARGET: 0x0051
```
──────────── trajectory.pl ────────────

The Perl program for this hybrid simulation is shown below and is straightforward. After instantiating an IO::HyCon object $ac in line 6 a readout group is defined in line 8. All elements of such a readout group can be read at once by the hybrid controller with minimal clock skew – something which is not a requirement in this particular case as the analog computer has been halted when the simulated shell hit ground and thus all variables are static.[184]

The following two calls to set_ic_time and set_op_time define these time intervals as 1 and 2 ms respectively, which is sufficient in this example with all integrators set to $k_0 = 10^3$.

The simulation starts with $\varphi = \frac{\pi}{4}$, set in line 12. Based on this value, $\dot{x}(0)$ and $\dot{y}(0)$ are computed and set in lines 14 and 15. Next, a single run consisting of an IC-period of 1 ms and an OP-phase of 2 ms is initiated by calling single_run_sync

---

[184] Except for some inevitable integrator drift in HALT mode.

in line 19. This method blocks further program execuation until either the defined OP-time has been reached or an external halt signal has occurred. In this simulation, the latter event will always precede a timeout since the shell will hit ground in less than 2 ms with $k_0 = 10^3$.

After each single run x_{shell} and x_{target} are read by invoking read_ro_group(). This returns a hash containing one key-value-pair for each element defined in the readout group. Based on these values δ is determined in line 22. This value is then used to determine a new elevation angle φ in line 23.[185] Prior to starting the next single run with this angle the current values of x_{shell}, x_{target}, and δ are printed on screen.

---------- trajectory.pl ----------
```perl
use strict;
use warnings;
use IO::HyCon;

my $ac = IO::HyCon->new();
$ac->set_ro_group('X', 'TARGET');
$ac->set_ic_time(1);
$ac->set_op_time(2);
$ac->enable_ext_halt();

my $phi = 3.1415 / 4; # Start with 45 degrees of elevation
while (1) {
    $ac->set_pt(0, cos($phi));
    $ac->set_pt(1, sin($phi));

    $ac->single_run_sync();
    my $result = $ac->read_ro_group();
    my $delta = $result->{X} - $result->{TARGET};
    $phi -= $delta / 10;
    printf("%+0.4f\t%+0.4f\t%+0.4f\n",

        $result->{X}, $result->{TARGET}, $delta);
}
```
---------- trajectory.pl ----------

---

[185] This is by no means the best way to determine the angle required to hit a defined target location. Its sole purpose is to serve as a programming example. If this were a simulation for an actual commercial or research project, a much more sophisticated and faster converging method of changing φ based on the values of δ would be employed.

7.4 Data gathering

Using a hybrid controller it is also possible to gather data either during HALT periods in a lenghty computer run or continuously during the OP period as shown in the following example.[186] The goal is to solve the one-dimensional wave equation

$$\frac{1}{c}\ddot{u} - \frac{\partial^2 u}{\partial x^2} = 0.$$

The derivatives with respect to x are approximated by a difference quotient

$$\frac{1}{c}\ddot{u} - \frac{u_{i-1} - 2u_i + u_{i+1}}{(\Delta x)^2} = 0$$

with $\Delta x = \frac{x}{n}$ and $n \in \mathbb{N}$. Defining

$$\lambda := \frac{c}{(\Delta x)^2} = \frac{cn^2}{x^2}$$

results in a form suitable to derive an analog computer setup:

$$\ddot{u}_i = \lambda \frac{u_{i-1} - 2u_i + u_{i+1}}{(\Delta x)^2}.$$

To simplify things further it is assumed that $n = 4$ and $\lambda = 1$ finally yielding the following set of coupled ordinary differential equations:

$$u_0 = \delta(t)$$
$$\ddot{u}_1 = u_0 - 2u_1 + u_2$$
$$\ddot{u}_2 = u_1 - 2u_2 + u_3$$
$$\ddot{u}_3 = u_2 - 2u_3 + u_4$$
$$\ddot{u}_4 = u_3 - 2u_4 + u_5$$
$$u_5 = 0$$

$\delta(t)$ represents an impulse occurring at the start of a simulation run, i.e., at $t = 0$. Figure 7.3 shows the resulting computer setup consisting of four basically identical cells, each containing two integrators, an inverting summer, and a coefficient potentiometer. Using the hybrid controller as a data logger it is now possible to gather the variables $u_1, \ldots, u_4$ during a simulation run.[187]

---

[186] Depending on the hybrid controller used, the amount of available memory for storing data may be rather limited. Sometimes it is an option to run one problem more than once, gathering one variable at a time instead of gathering a plethora of variables all at once if resolution time is not an issue.
[187] The sampling interval is automatically determined by the amount of free memory and the OP-time set for the run.

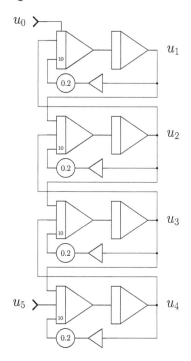

Fig. 7.3. Computer program for solving the one-dimensional wave equation

The corresponding configuration file for the hybrid controller looks basically like the minimal example shown on page 223 with the following changes and additions:

```
━━━━━━━━━━━━━━━━━━━━━━━━━━ wave_equation.yml ━━━━━━━━━━━
1  elements:
2      U1: 0260
3      U2: 0261
4      U3: 0262
5      U4: 0263
6  problem:
7      times:
8          ic: 10
9          op: 200
10     ro-group:
11         - U1
12         - U2
13         - U3
14         - U4
━━━━━━━━━━━━━━━━━━━━━━━━━━ wave_equation.yml ━━━━━━━━━━━
```

A problem-section has been added here. It contains the timing settings, the definition of a readout group (ro-group) and other optional subsections which are not required here. The readout group defines a list of computing elements which will be read during the OP phase of the simulation at equally-spaced intervals. In this case the readout group consists of the elements labeled U1, U2, U3, and U4.

The corresponding Perl program using this configuration file is shown below:

```perl
use strict;
use strict;
use warnings;
use IO::HyCon;

my $ac = IO::HyCon->new();   # Create a new hybrid controller object.
$ac->setup();                # Setup the analog computer
$ac->single_run_sync();      # Start a single run.
$ac->get_data();             # Get the data gathered
$ac->plot();                 # and generate a plot.
```

The method call $ac->setup() performs the setup of the analog computer based on the problem-section. If a readout group is defined, data will be gathered automatically by the hybrid controller during each run. The data gathered is stored in the $ac object and can be extracted by calling the method $ac->get_data(), which returns an array reference. Another useful method is $ac->plot(), which can be parameterized to yield different plotting styles and requires gnuplot being installed on the digital computer.

Calling $ac->plot() as in the example above yields the result shown in figure 7.4. The four lines in the graph represent the four variables $u_1, \ldots, u_4$. Each variable represents one location along the x-axis of the one-dimensional wave equation problem.

In problems like this it would be advantageous to get a more 3d-ish display of the variables being recorded. The plot() method supports such a display style, too. Calling $ac->plot(type => '3d') yields the plot shown in figure 7.5 which is much more intuitive than the overlaid 2d plots shown previously.[188]

---

[188] It should be noted that the sampling interval was about half as long as in the run yielding figure 7.4.

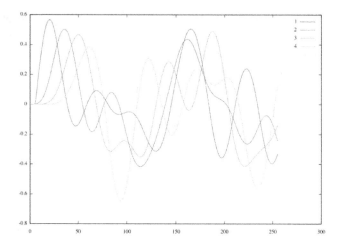

Fig. 7.4. Solution of the one-dimensional wave equation with the range x divided into four sections of equal width

7.5 Training an AI with an analog computer

The following example is more demanding with respect to the interplay between the digital and analog computers, which must work in parallel. The analog computer simulates an inverted pendulum, as described in section 6.33, while the digital computer implements a *reinforcement learning*[189] algorithm that learns to keep the inverted pendulum in its upright vertical position by actively balancing it.[190]

In order to learn how to balance the inverted pendulum the reinforcement learning system running on the digital computer reads four values from the analog computer: x and $\dot{x}$, describing the cart's position and velocity, as well as φ and $\dot{\varphi}$, representing the angle and angular velocity of the pendulum mounted on the cart. Since the setup described in section 6.33 only yields $\dot{\varphi}$ it is necessary to derive φ from its derivative by an additional integrator. The problem with a setup like this, containing two subprograms, both relying on $\dot{\varphi}$ and both employing integrators, is that these two groups of integrators will inevitably drift apart during prolonged simulation runs, due to unavoidable inaccuracies with the integrators' time scale

---

189 See [SUTTON et al. 2018] for an introduction to reinforcement learning.
190 A short video showing the overall hybrid computer setup and training results can be found here: https://www.youtube.com/watch?v=jDGLh8YWvNE, retrieved October 3rd, 2019. A more detailed description of the digital side of this simulation can be found at https://github.com/sy2002/ai-playground/tree/master/analog, retrieved October 3rd, 2019. The author would like to thank Mr. MIRKO HOLZER who implemented the AI-part of this hybrid computer setup. Cf. [HOLZER et al. 2020] for more details.

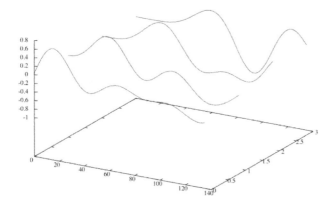

Fig. 7.5. Solution of the one-dimensional wave equation with the range x divided into four sections of equal width plotted with `type => '3d'`

factors. To avoid this problem the setup used for this hybrid simulation, shown in figure 7.6, employs two dedicated function generators, both using φ as their inputs.[191] Thus, φ can be derived from $\dot{\varphi}$ by a single integrator guaranteeing that all angle functions derived from it are consistent.

The reinforcement learning system running on the digital computer can push the cart on which the pendulum is mounted by applying a constant force to the left or right of the cart for a fixed time interval, like 10 or 20 ms. Figure 7.7 shows the subprogram implementing this application of force under program control. Using two digital outputs of the hybrid controller, D0 and D1, the cart can either be left uninfluenced (D0 = 0) or pushed (D0 = 1) to the left or right as determined by D1. An additional manually operated single pole, double throw switch allows the operator to unbalance the pendulum deliberately to see how well the reinforcment learning system can cope with unexpected perturbations.

Since the analog and digital computer work in parallel in this setup, it is of utmost importance that communication latencies are minimized. If a real-time operating system isn't used on the digital computer, the process running the reinforcement system should run with elevated priority and any additional computational or input/output load on the digital computer reduced as far as possible.

The reinforcement learning system has been implemented in Python 3.[192] The following skeleton listing shows the simple and self-explanatory main communi-

---

191 See appendix F for a schematic of these two function generators.
192 Its source code can be found at https://github.com/sy2002/ai-playground/blob/master/analog/, retrieved October 15th, 2019.

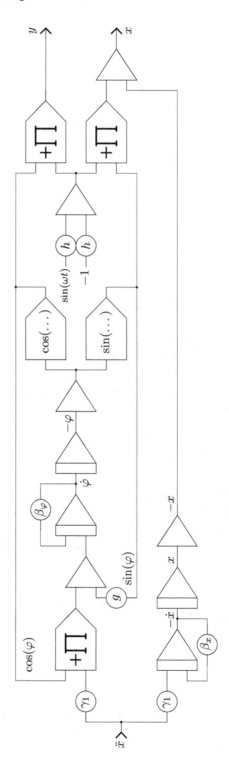

Fig. 7.6. Modified setup for the inverted pendulum

7.5 Training an AI with an analog computer — 233

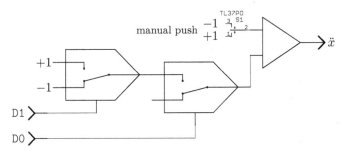

Fig. 7.7. Control circuit for the controlled inverted pendulum

cations routines implemented to send commands to the hybrid controller and to receive and interpret values read from computing elements:

```
                                 rl.py
1   import serial
2
3   from readchar import readchar
4   from time import sleep
5
6   # Hybrid Controller serial setup:
7   HC_PORT             = "/dev/cu.usbserial-DN050L10"
8   HC_BAUD             = 250000
9   HC_BYTE             = 8
10  HC_PARITY           = serial.PARITY_NONE
11  HC_STOP             = serial.STOPBITS_ONE
12  HC_RTSCTS           = False
13  HC_TIMEOUT          = 2
14
15  # Addresses of the computing elements to be read:
16  HC_SIM_X_POS        = "0223" # Position of the cart
17  HC_SIM_X_VEL        = "0222" # Velocity of the cart
18  HC_SIM_ANGLE        = "0161" # Angle of the pendulum
19  HC_SIM_ANGLE_VEL    = "0160" # Angular velocity of pendulum
20
21  HC_SIM_DIRECTION_1  = "D0" # Digital out 0=1: push right
22  HC_SIM_DIRECTION_0  = "d0" # Digital out 0=0: push left
23  HC_SIM_IMPULSE_1    = "D1" # Digital out 1=1: apply force
24  HC_SIM_IMPULSE_0    = "d1" # Digital out 1=0: stop applying force
25
26  # Model-1 Hybrid Controller: commands and responses:
27  HC_CMD_RESET        = "x"      # Reset hybrid controller
28  HC_CMD_INIT         = "i"      # Initial condition
```

```python
HC_CMD_OP              = "o"     # Start simulation run
HC_CMD_ROGROUP         = "f"     # Define a readout group
HC_RSP_RESET           = "RESET" # HC response to HC_CMD_RESET

# Send a command to the hybrid controller:
def hc_send(cmd):
    global dbg_last_sent
    dbg_last_sent = cmd
    hc_ser.write(cmd.encode("ASCII"))

# Get a response from the hybrid controller:
def hc_receive():
    # Since the hybrid controller terminates each output
    # with "\n", we can conveniently use readline.
    return hc_ser.readline().decode("ASCII").split("\n")[0]

# When reading values from the hybrid controller, the result
# is formatted as "<value><space><id/type>\n". The type
# information is ignored in the following:
def hc_res2float(str):
    f = 0
    try:
        f = float(str.split(" ")[0])
        return f
    except:
        print("ERROR #2: FLOAT CONVERSION:", str)
        print("Last command sent:", dbg_last_sent)
        print("Length of received string:", len(str))
        print("Hex output of string received:", ":" .
            join("{:02x}".format(ord(c)) for c in str))
        sys.exit(2)

# Query the current state of the simulation, which consists of
# the position and velocity of the cart as well as the the
# angle and angular velocity of the pendulum:
def hc_get_sim_state():
    hc_send(HC_CMD_ROGROUP)
    (res_x_pos, res_x_vel, res_angle, res_angle_vel) =
        hc_receive().split(';')
    return (hc_res2float(res_x_pos), hc_res2float(res_x_vel),
            hc_res2float(res_angle), hc_res2float(res_angle_vel))
```

7.5 Training an AI with an analog computer — 235

```python
71  # Reset the simulation (pendulum is in its upright position,
72  # cart is at x = 0):
73  def hc_reset_sim():
74      hc_send(HC_CMD_INIT)
75      sleep(0.05)          #time for capacitors to recharge
76      hc_send(HC_CMD_OP)
77      sleep(0.05)          #time for the HC to send the return string
78      hc_ser.flushInput()
79
80  # Push the cart to the left or right, depending on a:
81  def hc_influence_sim(a):
82      if (a == 1):
83          hc_send(HC_SIM_DIRECTION_1)
84      else:
85          hc_send(HC_SIM_DIRECTION_0)
86
87      hc_send(HC_SIM_IMPULSE_1)
88      sleep(HC_IMPULSE_DURATION / 1000.0)
89      hc_send(HC_SIM_IMPULSE_0)
90
91  # Initialize the serial line:
92  try:
93      hc_ser = serial.Serial(port=HC_PORT,
94                             baudrate=HC_BAUD,
95                             bytesize=HC_BYTE,
96                             parity=HC_PARITY,
97                             stopbits=HC_STOP,
98                             rtscts=HC_RTSCTS,
99                             dsrdtr=False,
100                            timeout=HC_TIMEOUT)
101     sleep(1.5)
102     dbg_last_sent = ""
103 except:
104     print("ERROR #1: SERIAL PORT CANNOT BE OPENED.")
105     sys.exit(1)
106
107 # Reset the hybrid controller:
108 received = ""
109 while (received != HC_RSP_RESET):
110     print("Hybrid Controller reset attempt...")
111     hc_send(HC_CMD_RESET)
112     sleep(1)
```

```
113        received = hc_receive()
114        if received != "":
115            print("  received:", received)
116
117    # Define the readout group for all four parameters:
118    hc_send('G' + HC_SIM_X_POS + ';' + HC_SIM_X_VEL + ';' +
119                 HC_SIM_ANGLE + ';' + HC_SIM_ANGLE_VEL + '.')
120
121    # Here follows the actual reinformcement learning code...
122    ...
```
──────────────── rl.py ────────────────

Using the analog computer as its sparring partner, the reinforcement learning systems learns to balance the inverted pendulum based on the values x, $\dot{x}$, φ, and $\dot{\varphi}$. Whenever φ exceeds about 30 degrees or the cart position x leaves the interval $[-1, 1]$ the current simulation run is terminated and the next run is started. At first, the reinforcement learning system acts rather randomly but improves quickly over time. At the end of the training the pendulum is held in a stable position more or less indefinitely.[193]

7.6 Hybrid solution of systems of linear equations

Using a hybrid computer it is also possible to solve systems of linear equations

$$\mathbf{A}\vec{x} = \vec{b} \qquad (7.1)$$

with an arbitrary degree of precision. This can be achieved by an iterative approach with an initial guess of the solution $\vec{x}_0$, which is refined repeatedly so that

$$\lim_{i \to \infty} \vec{x}_i = \vec{x}.$$

The iterative technique used in the following is the RICHARDSON *method*[194]

$$\vec{x}_{i+1} = \vec{x}_i - \omega \left(\mathbf{A}\vec{x}_i - \vec{b} \right), \qquad (7.2)$$

where $\omega = $ const is a *relaxation parameter*. Rearranging equation (7.2) yields

$$\vec{x}_{i+1} - \vec{x}_i = \omega \left(\vec{b} - \mathbf{A}\vec{x} \right).$$

---

[193] It would be simple to set up a *PID-controller* (a controller featuring a proportional, an integral, and a time derivative term) on the analog computer to stabilize the pendulum, but this would foil the aim of this experiment.
[194] See [RHEINBOLDT 2009] for details on this.

7.6 Hybrid solution of systems of linear equations

With $\omega \longrightarrow 0$ this transforms into the following system of differential equations

$$\dot{\vec{x}} = \vec{b} - \mathbf{A}\vec{x},$$

which has already been implemented in the program shown in figure 6.108 in section 6.31 with the little twist that $\vec{b}$ as well as $\mathbf{A}\vec{x}$ are multiplied with $\mathbf{A}^\mathrm{T}$ from the left, thus guaranteeing a symmetric and positive definite matrix and thus ensuring the convergence of the implementation

$$\dot{\vec{x}} = \mathbf{A}^\mathrm{T}\left(\vec{b} - \mathbf{A}\vec{x}\right).$$

In order to solve such a system on an analog computer $\mathbf{A}$ and $\vec{b}$ must be scaled by some factor $\lambda > 0$ which is possible since the equations are linear and

$$(\lambda \mathbf{A})\vec{x} = \lambda \vec{b} \iff \mathbf{A}\vec{x} = \vec{b}\ \forall \lambda > 0 \tag{7.3}$$

holds. Unfortunately it is not possible to scale the problem such that the analog computer will directly yield $\vec{x}$ since the solution is (obviously) not known in advance. Using a hybrid computer approach this problem can be turned into an advantage: Starting with an unscaled problem, the analog computer will typically quickly run into an overload which will be used to halt the computation by the hybrid controller. The result $\vec{x}_1$ obtained by this aborted run satisfies $\mathbf{A}\vec{x}_1 = \vec{b}_1 \neq \vec{b}$, which can be used to define a new system of linear equations

$$\mathbf{A}\vec{x}_2 = \vec{b}_2 = \vec{b} - \vec{b}_1$$

to be solved in the next iteration. If this could be solved without overload and to the desired precision it would yield

$$\mathbf{A}\left(\vec{x}_2 + \vec{x}_1\right) = \vec{b} - \vec{b}_1 + \vec{b}_1 = \vec{b},$$

so that $\vec{x} = \vec{x}_2 + \vec{x}_1$ would hold. $\vec{x}_2$ would thus "correct" the initial erroneous result $\vec{x}_1$ to yield the desired result. Since it is likely that this second step will also fail with an overload, this procedure is applied iteratively until the desired precision of the result is reached, i.e., until

$$\left|\left(\vec{b} - \mathbf{A}\vec{x}\right)_i\right| < \varepsilon$$

is satisfied for a given $\varepsilon > 0$ yielding

$$\vec{x} \approx \sum_{i=1}^{n} \vec{x}_i.$$

This method can be further refined to achieve an arbitrary degree of precision by rescaling the system of linear equations between successive iteration steps as shown in equation (7.3) with $\lambda_i > 1$ in order to make $\lambda_i \vec{b}$ larger and thus gain additional precision through the next $\vec{x}_i$. The final solution is then

$$\vec{x} \approx \sum_{i=1}^{n} \frac{\vec{x}_i}{\lambda_i}.$$

7.7 Solving PDEs with random walks

It has already been shown how partial differential equations can be solved by discretization along all but one of the variables involved. Unfortunately, this simple approach requires an immense number of computing elements rendering it unfeasible for problems of realistic size.

A completely different and interesting approach is the use of random walks[195] to solve partial differential equations, as the following example shows. The underlying theory is covered by the FEYNMAN-KAC *formula*, which links parabolic partial differential equations with stochastic processes. Of central importance here are WIENER *processes*, which are also known as BROWNIAN *motion*. The main requirement here is to have as many independent noise sources as there are spatial dimensions in the problem under consideration.[196]

The following example is mainly inspired by [SAWHNEY et al. 2020] and [SAWHNEY et al. 2022] and demonstrates how a partial differential equation of the form

$$\nabla(\alpha\nabla u) + \vec{\omega}\nabla u - \sigma u = -f \tag{7.4}$$

in a region Ω with boundary $\partial\Omega$ can be solved without resorting to creating a mesh covering Ω.[197]

This equation is a variant of a POISSON[198] equation. Here, f is a *source term* (describing an external heat source or sink, etc.). α is a *diffusion coefficient* modelling the rate of diffusion in the medium. $\vec{\omega}$ is a vector field called *drift coefficient* representing a motion of material on the region and σ represents an *absorption coefficient*.[199]

Here, a simplified variant of (7.4) namely $\Delta u - \sigma u = 0$ is solved over the region

$$\Omega = \left\{(x,y) \in \mathbb{R}^2 \mid |x| \leq 1 \wedge |y| \leq 1 \wedge (x,y) \notin B_r\right\}$$

with

$$B_r = \left\{(x,y) \in \mathbb{R}^2 \mid x^2 + y^2 \leq r^2\right\}$$

---

[195] See [HENZE 2018] for detailed information on random walks in general.
[196] The design and implementation of good noise sources with a bandwidth up to about 100 kHz is out of the scope of this book. [BENEKING 1971] gives a thorough account of the foundations of electronic noise in general.
[197] Typically, the construction of suitable meshes tends to get quickly rather complicated with increasing complexity of the structure.
[198] Named after its discoverer, SIMÉON DENIS POISSON.
[199] This is often called a *screening* coefficient. Accordingly, equation (7.4) is also called a *screened* POISSON equation.

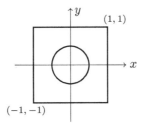

Fig. 7.8. Region Ω

denoting the two-dimensional ball around zero with radius r. This region is depicted in figure 7.8. The boundary is then

$$\partial \Omega = \partial B_r \cup \left\{ (x,y) \in \mathbb{R}^2 \,\Big|\, |x| = 1 \vee |y| = 1 \right\}$$

with

$$g(x,y) = \begin{cases} 2x & \text{if } (x,y) \in \partial B_r \\ \sin(2\pi x) & \text{if } |y| = 1 \\ \sin(2\pi y) & \text{if } |x| = 1, \end{cases}$$

describing the boundary conditions for the PDE, i.e., $u = g$ on $\partial \Omega$. u may be considered to represent a temperature within Ω with the boundary conditions representing external heat sources or sinks.

To get the desired the solution u at some coordinate (x, y) a number of independent random walks $X_i(t) = (x_i(t), y_i(t))$ are performed, starting at the location $X_i(0)$. The computing run lasts until such a random walk hits a boundary after a runtime of τ_i, i.e,. $X_i(\tau_i) \in \partial \Omega$. Due to the absorption σ, a boundary near to (x, y) has intuitively a bigger impact on the final value of u than one further away. This is taken into account by an exponential term $e^{-\sigma \tau}$ with τ representing the elapsed time until the random walk hits a boundary, yielding[200]

$$u = \mathbb{E}\left(e^{-\sigma \tau} g(X(\tau))\right) = \lim_{n \to \infty} \sum_{i=1}^{\infty} \frac{e^{-\sigma \tau_i} g(X_i(\tau_i))}{n}.$$

Figure 7.9 shows the analog computer setup for this problem. At the left are two independent noise sources[201], each of which is connected to a DC block as

---

[200] $\mathbb{E}(\dots)$ denotes the expected value.
[201] Used in this setup were two Wandel & Goltermann RG-1 noise sources yielding white noise signals ranging from DC to 100 kHz.

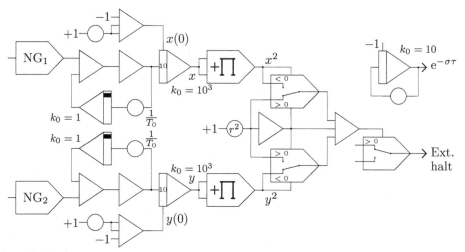

Fig. 7.9. Random walk program

shown in figure 5.14 in section 5.5.[202] The resulting values are then integrated yielding the x- and y-components of a single random walk.

Boundary checking consists of two parts: Detection of the outer boundary of Ω is done by halting the analog computer when an overload occurs, i.e., when x or y have crossed the boundary of the $[-1, -1] \times [1, 1]$ box. Detecting the boundary of B_r is more involved and is done by the program shown in the right half of figure 7.9, yielding an external halt signal which will also cause the computation run to halt.

In addition to the actual random walk, the program also implements the required exponential term by means of the integrator shown in the upper right.

The algorithm running on the attached digital computer is straightforward. First, the parameters σ, r^2, and $1/T_0$ are set. Then (x, y) are iterated over a set of points P_Ω for which the solution is required. For each such coordinate pair (x, y) the intial conditions $x(0)$ and $y(0)$ are set and an analog computer run is started. This run lasts until a halt is triggered by either an overload or the external halt signal. Depending on the values of x and y the boundary condition to be taken into account is then determined by a sequence of conditional statements, and u is updated accordingly.

Figure 7.10 shows a typical result gained by this method. The partition P_Ω equals an equidistant 50×50 grid, and for each point $(x, y) \in P_\Omega$ 200 random walks were performed.

---

[202] The two integrators forming these DC blocks are running continuously, i.e., not under the IC/OP control of the analog computer. If the filtering integrators were to be reset repeatedly, this would render the random signal unusable as the filter needs some time $\frac{1}{T_0}$ to reach a steady state.

7.7 Solving PDEs with random walks — 241

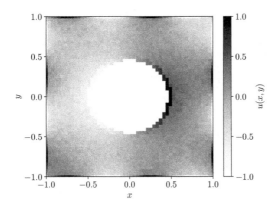

Fig. 7.10. Simulation result with a 50 × 50 grid over Ω

enable HALT on overload
enable HALT on external event
set σ
set r^2
set $\frac{1}{T_0}$
for all $(x, y) \in P_\Omega$ **do**
    $u \leftarrow 0$
    for $1 \leq i \leq n$ **do**
        set $x(0) = x$
        set $y(0) = y$
        run   ▷ Only HALT by overload or external event will terminate this
        if HALT on overload **then**   ▷ We hit the rectangular boundary
            if $|x| > 1$ **then**
                $g(x, y) \leftarrow \sin(2\pi y)$
            else   ▷ In this case $|y| > 1$
                $g(x, y) \leftarrow \sin(2\pi x)$
            end if
        else   ▷ The circle has been touched (external interrupt)
            $g(x, y) \leftarrow 2x$
        end if
        $u \leftarrow u + e^{-\sigma \tau} g(x, y)$
    end for
    $u(x, y) \leftarrow \frac{u}{N}$
end for
plot $u(x, y)$

8

Summary and outlook

Programming analog and hybrid computers is fundamentally different from the purely algorithmic approach used in digital computers, as the preceding chapters have shown. At first sight, using an analog computer seems to be much more involved and complicated than programming a digital computer. This is deceptive. Apart from the scaling issues, programming analog computers is much more straightforward than programming digital computers since the mathematical description of a problem is basically sufficient to set up an analog computer, while in the case of a digital computer a plethora of additional quirks and complications regarding the implementation on a specific machine have to be taken into account.

The main advantages of analog computers, either as stand-alone systems or as part of a hybrid computer setup, are

- the extremely high degree of parallelism that this class of machines exhibits,
- the high overall computational power, and
- the high energy efficiency.

Every change of a bit from 0 to 1 or vice versa in a digital computer requires a small parasitic capacitance to be charged or discharged. Together with other effects, such as the structure of typical output driver stages of logic gates, this results in a superlinear relationship between clock frequency and the overall energy consumption of a digital computer. Given today's clock frequencies of up to several GHz this results in power requirements of way over 100 W for a typical multi-core CPU chip. Apart from the challenge of cooling such devices, this is, quite simply, an excessive amount of energy for the computations and does not scale well in the long term.[203] Increasing the clock frequency of digital computers further to achieve higher computing power is not an easy feat since the problem of heat removal gets increasingly worse.

Digital computers are plagued by other problems as well – one of the most important effects is the upper bound of parallelism that can be obtained by an algorithmic approach for a certain problem, as described by AMDAHL's law. Only few problems are so well behaved that they can exploit the computing power of a highly-parallel supercomputer with thousands and sometimes several millions of individual cores. The majority of programs can't be scaled effectively over so many processing units.

Obviously, alternative approaches to classic digital computing are required to fulfill the ever increasing demands for computational power. Analog computers will be a central part of such novel systems, where they will act as coprocessors to speed up time-consuming or time-critical simulations, etc., as they do not suffer from the aforementioned problems.

Due to the representation of values as continuous voltages or currents and the continuous time of operation within an analog computer, all signals (with the exception of comparator and switch outputs) representing variables are "well behaved", i..e, they do not exhibit discontinuities. This results in a very high energy efficiency.

Abandoning the idea of a central memory system which stores commands and data to be processed in favor of interconnecting individual computing elements in order to set up an analog computer for a given problem, there is nothing like a *VON NEUMANN bottleneck*. Further, the abdication of an algorithm allows for a perfect degree of parallelism of the various computing elements within an analog computer.

Nevertheless, a lot of development work is required in the future to turn the classic analog computer with its cumbersome patch panel into a useful, general purpose, reconfigurable integrated circuit.[204] A main challenge will be the design and implementation of the configuration circuitry which will allow the elements of such an analog computer to be interconnected automatically under the control of a digital computer acting as a host system. In addition to the hardware, a software ecosystem must be developed, including a suitable (standardized) hardware description language to define the required interconnections of a particular analog

---

[203] It has been estimated that the world wide digital information technologies may account for 10 to 20% of the world's energy consumption in 2030. Even today the total energy consumption of these systems is well above 5% (see [GELENBE et al. 2015], [HEDDEGHEM et al. 2014], [JONES 2018], or [BELKHIR et al. 2018]).

[204] Currently quite a lot of research is being carried out in academia and industry to develop such ICs, but in 2023 there are no general purpose systems commercially available.

computer setup, libraries allowing the tight integration of the analog computer on a chip into a hybrid computer are required, etc.

These challenges are formidable, but the possibilities of analog computing are nearly endless and will more than justify all the necessary research and development expenses.

Given the high energy efficiency of analog computers compared with digital approaches, they will create new fields of application – such as medical implants, which might consume so little electrical energy that they could be powered by energy harvesting within the body, thus making energy storage devices that must be changed or recharged at regular intervals superfluous. It will also find numerous applications in mobile and wearable devices for signal pre- and post-processing to save precious battery capacity. Even complex tasks such as trigger word detection for assistance systems can be implemented using analog computer techniques, saving power in standby mode, etc.

Other applications will benefit from the fact that a system that is not controlled by a program stored in some volatile memory cannot be "hacked" in a traditional sense.[205]. This will make analog computers the systems of choice when it comes to control systems in critical infrastructure systems, systems on which human lives depend, etc.

Also, most branches of *artificial intelligence*, *(AI)*, and *machine learning* will benefit substantially from the application of analog computing techniques. Spiking and bursting neurons can be directly implemented as analog circuits and the fact that neuronal networks are basically defined by their interconnection structure, together with their synaptic weights, fits perfectly with the paradigm of analog computing.

All in all, analog computing is an extermely promising approach for future high performance and/or low energy computing and will be part of most future computing systems.

Happy analog computing!

---

205 If a reconfigurable analog computer is used, the reconfiguration capability must be physically disabled to gain that benefit.

The Laplace transform

One of the standard tools for solving differential equations analytically is the LAPLACE *transform*, a parameter integral named after PIERRE SIMON LAPLACE. Although this is not directly connected to analog computer programming, the basic ideas and techniques are quite interesting and can sometimes be used to cross check the analog computer solution of a problem.[206]

It is defined as

$$F(s) = \mathcal{L}\left(f(t)\right) = \int_0^\infty f(t) e^{-st} \, dt \quad (A.1)$$

with $s \in \mathbb{C}$ and transforms differential equations into algebraic equations which can typically be solved much more easily. The result obtained is then transformed back by means of an inverse transform $\mathcal{L}^{-1}\left(F(s)\right)$, which is normally done by using tabulated functions.

A.1 Basic functions

This section presents the derivation of the LAPLACE transforms for some basic functions.

---

[206] Further information on this particular transform can be found in [WIDDER 2010]. Three wonderful historic texts on the subject are [CARSLAW et al. 1941], [VAN DER POL 1987], and [DOETSCH 1970]. Its particular application to problems in engineering and especially control theory can be found in many texts such as [SENSICLE 1968] and [FÖLLINGER et al. 2021]. [DUFFY 1994] shows advanced transforms for the solution of partial differential equations.

A.1.1 Step function

One of the most basic functions is the *step function*

$$f(t) = \begin{cases} 0 & t < 0 \\ a & t > 0 \end{cases}$$

with amplitude $a \in \mathbb{R}^+$. In the case of $a = 1$ it is called *unit step function*, denoted $u(t)$ which is also known as HEAVISIDE *step function* after OLIVER HEAVISIDE.[207] Its LAPLACE transform is[208]

$$\mathcal{L}(u(t)) = \int_0^\infty a e^{-st}\, dt = a \int_0^\infty e^{-st}\, dt. \tag{A.2}$$

Substituting $\varphi = -st$ yields $dt = -\frac{1}{s}\, d\varphi$, so (A.2) simplifies to

$$\mathcal{L}(u(t)) = -\frac{a}{s} \int_0^\infty e^\varphi\, d\varphi = -\frac{a}{s} \left[e^{-st} \right]_0^\infty = \frac{a}{s}. \tag{A.3}$$

A variant of the step function is the *delayed unit step function*

$$u(t - T) = \begin{cases} 0 & t < T \\ 1 & t > T \end{cases}$$

with the following LAPLACE transform

$$\begin{aligned}
\mathcal{L}(u(t-T)) &= \int_0^\infty u(t-T) e^{-st}\, dt \\
&= \int_0^T u(t-T) e^{-st}\, dt + \int_T^\infty u(t-T) e^{-st}\, dt \\
&= \int_T^\infty e^{-st}\, dt = -\frac{1}{s} \left[e^{-st} \right]_T^\infty = \frac{e^{-sT}}{s}.
\end{aligned}$$

---

[207] See [YAVETZ 1995] for more information on HEAVISIDE's work.
[208] Note the lower bound of integration.

A.1.2 Delta function

In many cases, the DIRAC *delta function*[209] $\delta(t)$ is used as input for a dynamic system.[210] It has zero width but satisfies

$$\int_{-\infty}^{\infty} \delta(t)\,\mathrm{d}t = 1$$

and is therefore often called the *unit impulse*. It can be understood as the derivative of the unit step function $u(t)$. Its LAPLACE transform can therefore be derived by integrating by parts as follows:

$$\mathcal{L}(\delta(t)) = \int_0^{\infty} \left(\frac{\mathrm{d}}{\mathrm{d}t}u(t)\right) e^{-st}\,\mathrm{d}t$$

$$= \left[u(t)e^{-st}\right]_0^{\infty} - \int_0^{\infty} u(t)\left(-se^{-st}\right)\mathrm{d}t$$

$$= 0 + s\int_0^{\infty} e^{-st}\,\mathrm{d}t$$

Substituting $\varphi = -st$ as before yields $\mathrm{d}t = -\frac{1}{s}\mathrm{d}\varphi$ so that

$$s\int_0^{\infty} e^{-st}\,\mathrm{d}t = -\int_0^{\infty} e^{\varphi}\,\mathrm{d}\varphi = \left[e^{-st}\right]_0^{\infty} = 1.$$

Thus, $\mathcal{L}(\delta(t)) = 1$.

A.1.3 Ramp function

Another basic function is the *ramp function*

$$f(t) = \begin{cases} 0 & t < 0 \\ at & t > 0 \end{cases}.$$

Its LAPLACE transform can be derived as follows, using integration by parts:

$$\mathcal{L}(f(t)) = a\int_0^{\infty} te^{-st}\,\mathrm{d}t = a\left(\left[-\frac{te^{-st}}{s}\right]_0^{\infty} - \int_0^{\infty} -\frac{e^{-st}}{s}\,\mathrm{d}t\right) \quad (\text{A.4})$$

---

209 Named after PAUL ADRIEN MAURICE DIRAC.
210 $\delta(t-T)$ typically denotes the *delayed delta function*.

Substituting $\varphi = -st$ a second time gives

$$-\int_0^\infty \frac{e^{-st}}{s}\,dt = \frac{1}{s^2}\int_0^\infty e^\varphi\,d\varphi = \left[\frac{e^{-st}}{s^2}\right]_0^\infty \quad (A.5)$$

Combining (A.4) and (A.5) results in[211]

$$\mathcal{L}(f(t)) = a\left[-\frac{te^{-st}}{s} - \frac{e^{-st}}{s^2}\right]_0^\infty = a\left[-\frac{(st+1)e^{-st}}{s^2}\right]_0^\infty = \frac{a}{s^2}.$$

A.1.4 Exponential and trigonometric functions

The LAPLACE transform of the exponential function has the form

$$\mathcal{L}(e^{-at}) = \int_0^\infty e^{-at}e^{-st}\,dt = \left[\frac{e^{-(s+a)t}}{-(s+a)}\right]_0^\infty = \frac{1}{s+a}.$$

The basic trigonometric functions can also be transformed in a straightforward fashion using the identities

$$\sin(\varphi) = \frac{e^{i\varphi} - e^{-i\varphi}}{2i} \quad \text{and} \quad (A.6)$$

$$\cos(\varphi) = \frac{e^{i\varphi} + e^{-i\varphi}}{2}. \quad (A.7)$$

Using (A.6) gives

$$\mathcal{L}(\sin(\omega t)) = \int_0^\infty \sin(\omega t)e^{-st}\,dt = \int_0^\infty \left(\frac{e^{i\omega t} - e^{-i\omega t}}{2i}\right)e^{-st}\,dt$$

$$= \frac{1}{2i}\int_0^\infty \left(e^{(i\omega-s)t} - e^{-(i\omega+s)t}\right)dt = \frac{1}{2i}\left[\frac{e^{(i\omega-s)t}}{i\omega - s}\right]_0^\infty - \frac{1}{2i}\left[\frac{e^{-(i\omega+s)t}}{-i\omega - s}\right]_0^\infty$$

$$= \frac{1}{2i}\left(\frac{1}{-s+i\omega} - \frac{1}{-s-i\omega}\right) = \frac{\omega}{s^2+\omega^2}.$$

Using (A.7) the relation

$$\mathcal{L}(\cos(\omega t)) = \frac{s}{s^2+\omega^2}$$

can be derived analogously.

---

[211] It can be shown that $\mathcal{L}(at^n) = \frac{an!}{s^{n+1}}$.

A.2 Laplace transforms of basic operations

Things really get interesting when the LAPLACE transforms of operations instead of simple functions are computed. This is shown below for the time derivative and the time integral of a function. Let

$$f(t) = \frac{dx}{dt},$$

then integration by parts gives

$$\mathcal{L}(f(t)) = \int_0^\infty \frac{dx}{dt} e^{-st}\, dt$$

$$= \left[xe^{-st} \right]_0^\infty - \int_0^\infty -sxe^{-st}\, dt$$

$$= s\mathcal{L}(x) - x(0).$$

Here $x(0)$ denotes the initial value of x at $t = 0$. The interesting fact is that the time derivative of a function $f(t)$ can be transformed into the much simpler multiplication of the LAPLACE transform of the function with s.

This can be extended to higher derivatives as well. From

$$f(t) = \frac{d^2x}{dt^2}$$

follows that

$$\mathcal{L}(f(t)) = \int_0^\infty \frac{d^2x}{dt^2} e^{-st}\, dt = s^2 \mathcal{L}(f(t)) - sx(0) - \left[\frac{dx}{dt} \right]_{t=0}$$

and so forth.[212]

If multiplication with s is the transform of a time derivative, it is tempting to assume that integration will be transformed into a division by s. This, indeed, is the case. Let

$$f(t) = \int_0^t x\, dt + c$$

with c being a constant real value. Its LAPLACE transform is

$$\mathcal{L}(f(t)) = \int_0^\infty \left(\int_0^t x\, dt \right) e^{-st}\, dt + \frac{c}{s}$$

---

[212] $\left[\frac{dx}{dt} \right]_{t=0}$ will often be written as $[\dot{x}]_{t=0}$ for brevity.

$$= \left[-\frac{1}{s}e^{-st}\int_0^t x\,dt\right]_0^\infty + \frac{1}{s}\int_0^\infty xe^{-st}\,dt + \frac{c}{s}$$

$$= \frac{1}{s}\mathcal{L}(f(t)) + \frac{c}{s}.$$

The constant $\frac{c}{s}$ is the initial value of the integral.

A.3 Further characteristics

Given two functions $f(t)$ and $g(t)$ which can be subjected to a LAPLACE transform yielding $F(s)$ and $G(s)$, the transform has the following useful characteristics:

$$\mathcal{L}(\lambda f(t) + g(t)) = \lambda F(s) + G(s) \qquad \text{(linearity)}$$

$$\mathcal{L}(f(\lambda t)) = \frac{1}{\lambda}F\left(\frac{s}{\lambda}\right) \qquad \text{(scale factor)}$$

$$\mathcal{L}\left(e^{-\beta t}f(t)\right) = F(s+\beta) \qquad \text{(exponential damping)}$$

$$\mathcal{L}\left(f(t-T)u(t-T)\right) = e^{-Ts}F(s) \qquad \text{(time shift)}$$

$$\mathcal{L}\left((f\star g)(t)\right) = F(s)G(s) \qquad \text{(convolution)}$$

A.4 Inverse Laplace transform

The inverse LAPLACE transform is defined as

$$f(t) = \mathcal{L}^{-1}(F(s)) = \frac{1}{2\pi i}\lim_{T\to\infty}\int_{\gamma-iT}^{\gamma+iT} e^{st}F(s)\,ds$$

with $y \in \mathbb{R}$ denoting a vertical line in the complex number plane. y is chosen so that all singularities of $F(s)$ are on its left. This transform is known as the BROMWICH integral, after THOMAS JOHN L'ANSON BROMWICH and is typically solved using the *residue theorem*.[213]

---

[213] The line integral of an analytic function $f(z)$ over a closed curve γ can be determined by

$$\oint_\gamma f(z)\,dz = 2\pi i \sum_{j=1}^n \text{I}(\gamma, a_j)\text{Res}(f, a_j).$$

Here, the a_j denote a finite set of points where $f(z)$ has singularities while $\text{I}(a_j)$ is the *winding number* of the path γ around the point a_j and $\text{Res}(f, a_j)$ is the residue of $f(z)$ at the point a_j. See [CARSLAW et al. 1941], [DUFFY 1994], etc., for more information on this technique.

Normally, this process is avoided by using tables of functions and their LAPLACE transforms, e. g.:

$$\mathcal{L}\left(\delta(t)\right) = 1$$
$$\mathcal{L}\left(\delta(t-T)\right) = e^{-Ts}$$
$$\mathcal{L}\left(u(t)\right) = \frac{1}{s}$$
$$\mathcal{L}\left(\frac{1}{\sqrt{\pi t}}\right) = \frac{1}{\sqrt{s}}$$
$$\mathcal{L}\left(t^n\right) = \frac{n!}{s^{n+1}}$$
$$\mathcal{L}\left(t^n e^{-at}\right) = \frac{n!}{(s+a)^{(n+1)}}$$
$$\mathcal{L}\left(\frac{1}{a}\left(1 - (1+at)e^{-at}\right)\right) = \frac{1}{s(s+a)^2}$$
$$\mathcal{L}\left(\frac{e^{-at} - e^{-bt}}{b-a}\right) = \frac{1}{(s+a)(s+b)}$$
$$\mathcal{L}\left(\sin(\omega t)\right) = \frac{\omega}{s^2+\omega^2}$$
$$\mathcal{L}\left(\cos(\omega t)\right) = \frac{s}{s^2+\omega^2}$$
$$\mathcal{L}\left(e^{-at}\sin(\omega t)\right) = \frac{\omega}{(s+a)^2+\omega^2}$$
$$\mathcal{L}\left(e^{-at}\cos(\omega t)\right) = \frac{s+a}{(s+a)^2+\omega^2}$$
$$\mathcal{L}\left(\sinh(\omega t)\right) = \frac{\omega}{s^2-\omega^2}$$
$$\mathcal{L}\left(\cosh(\omega t)\right) = \frac{s}{s^2-\omega^2}$$
$$\mathcal{L}\left(\frac{\sqrt{a^2+\omega^2}}{\omega}\sin\left(\omega t + \tan^{-1}\left(\frac{\omega}{a}\right)\right)\right) = \frac{s+a}{s^2+\omega^2} \quad (A.8)$$
$$\mathcal{L}\left(\frac{\sin(\omega t)}{t}\right) = \arctan\left(\frac{\omega}{t}\right)$$

A.5 Example

Solving a differential equation such as

$$\ddot{x} + x = \delta(t)$$

using the LAPLACE transform is pretty straightforward. First, the transform is applied to both sides yielding

$$s^2 \mathcal{L}(x) - sx(0) - [\dot{x}]_{t=0} + \mathcal{L}(x) = 1.$$

The initial conditions $x(0) = 1$ and $[\dot{x}]_{t=0} = 0$ give

$$s^2 \mathcal{L}(x) - s + \mathcal{L}(x) = 1,$$

which can be rearranged into

$$(s^2 + 1)\mathcal{L}(x) = s + 1$$

and finally into

$$\mathcal{L}(x) = \frac{s+1}{s^2+1}.$$

Instead of computing the inverse transform, the table of typical LAPLACE transforms is used to determine

$$x = \mathcal{L}^{-1}\left(\frac{s+1}{s^2+1}\right),$$

the right-hand side of which matches equation (A.8) in the preceding section with $a = 1$ and $\omega^2 = 1$ yielding the final solution of this differential equation:

$$x = \sqrt{2}\sin\left(t + \frac{\pi}{4}\right)$$

A.6 Block diagrams and transfers functions

In engineering, dynamic systems are often described as *block diagrams* showing all forward and feedback paths of the system under consideration, such as transducers, amplifiers, servomotors, etc. The individual blocks are described by a *transfer function* Φ which defines how the input and output of the block are linked together:

$$\frac{\text{output}}{\text{input}} = \Phi.$$

These transfer functions are typically written as LAPLACE transforms.[214] They can be simulated by either passive electronic elements alone, a combination of operational amplifiers and passive elements, or by analog computing elements. Using tables linking typical transfer functions with their corresponding implementation

---

[214] Instead of the LAPLACE transform variable s, which is identified with differentiation and integration in the case of $\frac{1}{s}$, the classic p operator is often used:

$$p = \frac{d}{dt}, \frac{1}{p} = \int_0^t dt$$

using analog computing elements, simulations of complex systems described by block diagrams and corresponding transfer functions can easily be set up.

Details of this technique, which exceeds the scope of this book, can be found in [SENSICLE 1968], [JOHNSON 1956, p. 45 et seq.], [GILOI et al. 1963, p. 288 et seq.], [CARLSON et al. 1967, p. 191 et seq.], [SYDOW 1964, p. 128 et seq.], and other classic text books. Good examples of application are described in [WEEDY et al. 1998] and [GILES 1976].

B

Solving the heat equation with a passive network

The two-dimensional heat equation as treated in section 6.30 can also be solved by means of a two-dimensional grid consisting of a number of identical passive resistor/capacitor (RC) nodes as shown below.[215] The aim is to model the heat flow in a two-dimensional sheet of thermally conductive material such as a thin metal plate. The flow is described by

$$\dot{u} = \alpha \nabla^2 u. \tag{B.1}$$

To model the behavior of this equation by means of a discrete two-dimensional grid of passive elements, it is approximated by finite differences yielding

$$\dot{u}_{i,j} = \alpha \left(u_{i-1,j} + u_{i+1,j} + u_{i,j-1} + u_{i,j+1} - 4u_{i,j} \right) + q_{i,j}$$

with $u_{i,j}$ denoting a single grid node, and $q_{i,j}$ representing a heat source or sink connected to this node. Such a node is shown in figure B.1 and consists of four resistors and a capacitor with a common junction. Applying KIRCHHOFF's current law to the center node $u_{i,j}$ yields the following expression for the voltage at this node:

$$\frac{(u_{i,j-1} - u_{i,j}) + (u_{i,j+1} - u_{i,j}) + (u_{i-1,j} - u_{i,j}) + (u_{i+1,j} - u_{i,j})}{R} - C\dot{u}_{i,j} + I_{i,j} = 0.$$

---

[215] The author would like to thank Dr. CHRIS GILES for his invaluable support in setting up this example.

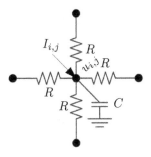

Fig. B.1. A single RC-node

The terms in the enumerator are the currents flowing through the resistors, the second term describes the current through the capacitor, and the last term is an optional current that may be injected into this node. Rearranging then yields

$$C\dot{u}_{i,j} = \frac{(u_{i,j-1} - u_{i,j}) + (u_{i,j+1} - u_{i,j}) + (u_{i-1,j} - u_{i,j}) + (u_{i+1,j} - u_{i,j})}{R} + I_{i,j}. \tag{B.2}$$

Equations (B.1) and (B.2) have the same structure so that a grid consisting of many nodes as shown in figure B.1 may be used to solve the two-dimensional heat equation problem[216] by considering

$$\text{voltage} \equiv \text{temperature and}$$
$$\text{current} \equiv \text{heat flux}.$$

In the following, a plate of dimension 16×16 cm^2 as shown in figure B.2 is considered. Its top and bottom surfaces are perfectly insulated while its outside boundary is held at a constant temperature ($\equiv$ voltage). The plate is divided into a 1×1 cm^2 grid. Exploiting the symmetries along the x- and y-axes, it is sufficient to model only one quadrant of the plate.[217]

The upper right quadrant is thus modelled by a passive 8×8 grid consisting of RC combinations as shown in figure B.3. The connection labeled $I_{0,0}$ at the lower left corner allows to apply a current (heat flux) or a voltage (temperature) to the corner node $u_{0,0}$. The connections α and β implement the fixed boundary values along the edge of the quadrant. The physical implementation of this grid is shown in figure B.4. It was built using 180k resistors and 100nF capacitors.

---

[216] Approaches like this were in widespread use well into the 1960s for solving a wide range of heat transfer, electromagnetic, and fluid flow problems.

[217] The symmetry along the 45 degree diagonals can not be as easily exploited using a passive model like this as it was in the example shown in section 6.30.

B Solving the heat equation with a passive network

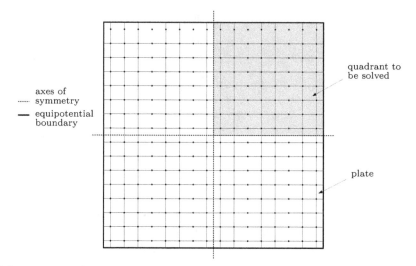

Fig. B.2. 16 × 16 cm plate structure

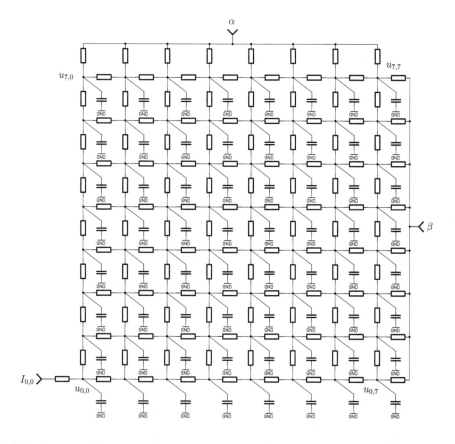

Fig. B.3. Passive network for one quadrant

Fig. B.4. Implementation of the two-dimensional RC-network

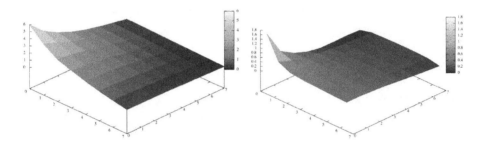

Fig. B.5. Steady state of the network and differences between adjacent nodes

The two graphs shown in figure B.5 depict the steady state after forcing the temperature at node $u_{0,0}$ to a fixed value. The left hand picture shows the temperature distribution across the plate quadrant. The graph on the right shows the heat flux densities, which were obtained by measuring the voltages across the individual resistors which correspond to the heat flux between adjacent nodes. This yields the x- and y-components of the heat flux which allowed the heat flux magnitudes in between the nodes to be calculated.

Figure B.6 shows the time dependent voltages at the diagonal nodes $u_{i,i}$ after injecting a pulse $I_{0,0}$ at the corner grid node $u_{0,0}$. The diffusion of the temperature across the plate quadrant can clearly be seen as well as the eventual settling of the nodes along the diagonal to the boundary condition temperature as defined by the inputs α and β.

As inflexible as passive networks like this are with regard to their spatial structure, they can be used to solve partial differential equations of a given structure with varying boundary conditions very efficiently. [VOLYNSKII et al. 1965] contains a lot of practical examples and theoretical background.

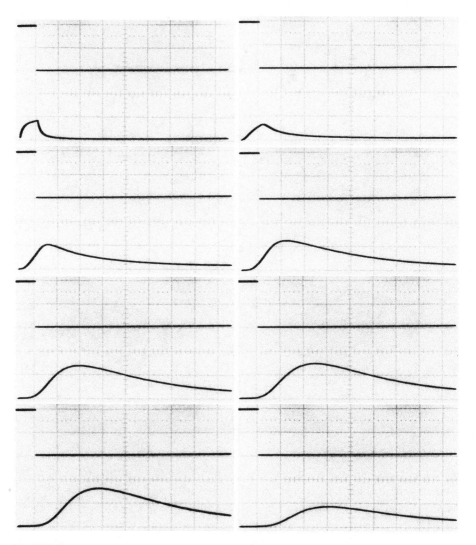

Fig. B.6. Response to a step input at nodes $u_{0,0}, u_{1,1}, \ldots u_{7,7}$ (from left to right and up/down)

A simple hybrid controller for THE ANALOG THING

This chapter describes a simple Arduino®[218] based hybrid controller for THE ANALOG THING.[219] At its heart is an Arduino® Mega 2650 single board microcontroller, which is connected to the HYBRID connector of the analog computer. The microcontroller can take over control of the analog computer, i.e., it can control its mode of operation (IC, OP, or HALT), and it can sample data from up to four analog outputs.[220]

Table C.1 shows the connections between this connector and the Arduino® input/output pins.[221] These connections allow the Arduino® to control the operation of the analog computer and to gather data from the four analog outputs. The hybrid controller software[222] implements the following commands:

---

[218] See https://www.arduino.cc, retrieved on December 1st, 2022.
[219] The source code for this device can be found at https://github.com/anabrid/hardware/tree/main/the-analog-thing/arduino_2650_hybrid_controller, retrieved on December 1st, 2022.
[220] With a minimal setup like this it is not possible to automatically set coefficients. This would require additional digital potentiometers such as the AD5293 or multiplying DACs connected to the Arduino®.
[221] The analog outputs of THE ANALOG THING available on the HYBRID connector correspond to the values patched to the X-, Y-, Z-, and U-jacks on the patch panel. These signals are fed to voltage dividers and level shifters yielding analog signals between 0 and 3.3 V, which can be fed directly to the analog inputs of most modern microcontrollers.
[222] See https://github.com/anabrid/hardware/tree/main/the-analog-thing/arduino_2650_hybrid_controller/simple_hybrid_controller, retrieved December 1st, 2022.

HYBRID pin	Name	Description	Destination on the Arduino® board
1	IN-X	analog x-input	n. c.
2	HYB-X	analog x-output	AnalogIn 0
3	IN-Y	analog y-input	n. c.
4	HYB-Y	analog y-output	AnalogIn 1
5	IN-Z	analog z-input	n. c.
6	HYB-Z	analog z-output	AnalogIn 2
7	IN-U	analog u-input	n. c.
8	HYB-U	analog u-output	AnalogIn 3
9,10	GND	Ground	GND
11,12	VUSB	+5 V	n. c.
13	DIR	Enable hybrid mode	D2
14	ModeOP	Integrator control	D3
15	Voffset	Offset voltage output	n. c.
16	ModeIC	Integrator control	D4

Table C.1. Mapping of the HYBRID port on THE ANALOG THING to the Arduino® Mega 2650 input/output pins

arm: Arm the data logger for data capturing to start at the begin of the next single run.

channels=<value>: This command sets the number of channels that are to be logged. <value> can be set to 1, 2, 3, or 4.

disable: Disable the hybrid controller.[223] This allows THE ANALOG THING to work as a standalone system without the hybrid controller interfering.

enable: Enable the hybrid controller, which then takes over control of the attached analog computer.

halt: Set the analog computer to HALT-mode.

help: Print a short help text listing all of the available commands.

ic: Set the analog computer to IC-mode.

ictime=<value>: Set the initial condition time to <value> milliseconds. If any of the integrators of the analog computer are configured to run in SLOW mode (corresponding to $k_0 = 10$), this interval should be at least several ten milliseconds long.

interval=<value>: This sets the data sampling interval to <value> milliseconds. Data sampled will be transmitted through the USB-connection of the microcontroller with individual channel values separated by a semicolon.

op: Set the analog computer to OP-mode.

---

[223] At power on the hybrid controller will be in this state.

optime=<value>: Set the operation mode time to <value> milliseconds. This setting, together with ictime, controls the behavior of repetitive operation or single run.

rep: Start repetitive operation. The attached analog computer cycles between IC- and OP-mode in this mode until switched explicitly to IC, OP, or HALT mode by the hybrid controller.

run: Start a single run on the analog computer.

status: Return general status information about the configuration of the hybrid controller.

Figure C.1 shows a typical setup consisting of THE ANALOG THING and an attached microcontroller board. A typical command sequence for performing a single run of a program gathering one channel of data would look like this:

```
ictime=50
optime=3000
interval=5
channels=1
arm
enable
run
```

The setup shown in figure C.1 implements the HINDMARSH-ROSE model.[224] The above command sequence yields a data set like the one shown in figure C.2.[225] The seemingly varying pulse heights are caused by the rather long sample interval, a shorter interval should be preferred.

---

[224] See section 6.26.

[225] The plot was created with gnuplot, see http://www.gnuplot.info, retrieved December 1st, 2022.

264 —— C A simple hybrid controller for THE ANALOG THING

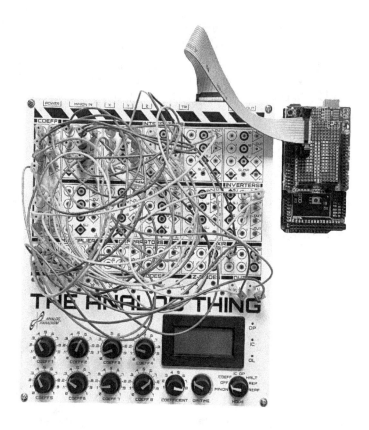

Fig. C.1. THE ANALOG THING controlled by an Arduino®Mega 2650

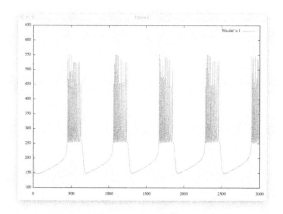

Fig. C.2. Result from a three second run of the Hindmarsh-Rose model

An oscilloscope multiplexer

Most oscilloscopes provide at least two y-channels which allow the display of two curves at once on the screen. Although this is often sufficient for work with an analog computer, sometimes it is desirable to have not only two y-inputs but several pairs of x- and y-inputs. This allows multiple figures, each defined by a corresponding x, y signal pair, to be displayed. Since this is not possible with most standard oscilloscopes the need for an x, y oscilloscope multiplexer arises.[226]

Figure D.1 shows the prototype implementation of the four channel multiplexer described in the following.

The schematic is shown in figure D.2. It features four y- and four x-inputs and yields a single y- and x-output to the oscilloscope. In addition to this, a z-signal is generated which can be used for automatic beam blanking if the oscilloscope supports an external beam control input.

The lower third of the figure contains the clock circuitry which can be run in either of two modes as determined by the position of switch S1. In its top position the multiplexer is free running controlled by a simple astable multivibrator based on a NE555 (IC1). If S1 is set to the lower position, the clock signal is activated every time the analog computer is in OP-mode. This is especially useful when repetitive operation is selected as the oscilloscope will switch from one input pair to the next in conjunction with the operation cycles. This circuit also generates the z-signal which turns the beam off whenever the analog computer is not in OP-mode as well as while switching from one input channel to another. The operational amplifier IC3 in conjunction with the potentiometer R5 is used to adjust the blanking signal level between $+15$ V and -15 V to suit the particular oscilloscope.

---

[226] If this functionality is only rarely used, the electronic switches of an analog computer can be used in pairs with proper control logic.

D An oscilloscope multiplexer

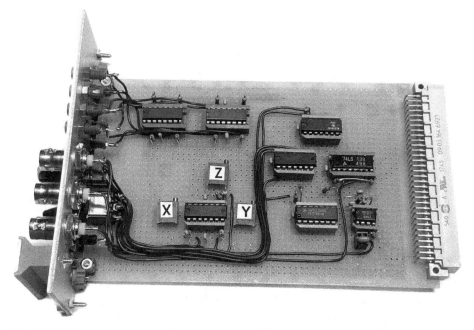

Fig. D.1. Implementation of the multiplexer

The middle section of the schematic shows the two bit counter which selects one out of the four input signal pairs. The counter itself is built from two JK-flip-flops (IC6). Using switch S2 the number of channels displayed can be selected from one to four. The two address bits feed a two-to-four demultiplexer (IC4B) which yields four active-low logic signals controlling the two electronic quad switches IC7 and IC9. Both outputs to the oscilloscope are buffered by simple impedance converters (IC8 and IC10).

D An oscilloscope multiplexer — 267

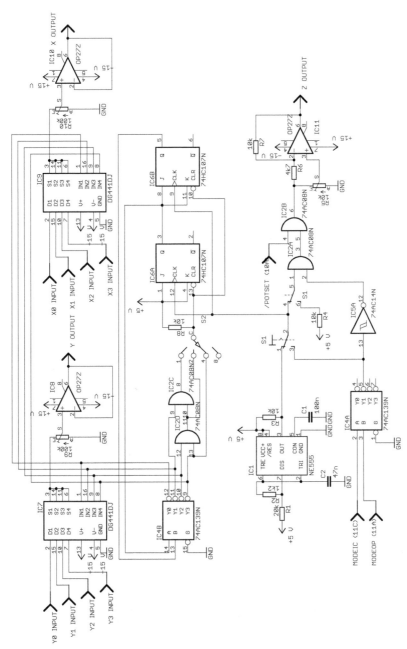

Fig. D.2. Four channel x, y oscilloscope multiplexer

A log() function generator

The logarithm is a particularly useful function as has been seen in section 6.28 where this function was required for the simulation of an air flow. The literature abounds with electronic logarithm function generators – either based on polygon approximation, see section 2.6, or using a suitable PN-junction in the feedback path of an operational amplifier. Both approaches are quite tricky to implement due to the non-negligible temperature coefficients of the diodes and transistors required. Fortunately, there exists an integrated circuit, the LOG112, which implements a high-precision logarithm function generator that can be easily interfaced to an analog computer. The behavior of this device is characterized by

$$V_{\text{out}} = \frac{1}{2} \log_{10}\left(\frac{I_{\text{in}}}{I_{\text{ref}}}\right)\left(1 + \frac{R3 + R4}{R2}\right)$$

where I_{in} and I_{ref} denote an input and a reference current.

Figure E.1 shows the circuit of the overall computing element: The reference current I_{ref} is defined by R5 which should have a precision of at least 0.05%.[227] The input current I_{in} is defined by R1 which should have the same precision.

The input/output voltage relation of this circuit is shown in table E.1. Input values less than 0.001 V will result in an overload that could be detected either by means of a analog computer comparator or by a dedicated detector circuit based on an LM339 or the like. The values of the table should be used to calibrate the circuit by setting the potentiometer R4 accordingly.

---

[227] This resistor can also be implemented by paralleling four high-precision 1M resistors if 250k are not readily available.

E A log() function generator

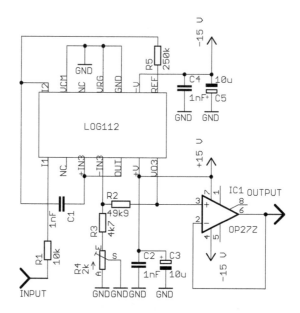

Fig. E.1. log() function generator

Exponent	Input	Output
−3	0.001 V	−10 V
−2	0.01 V	−5 V
−1	0.1 V	0 V
0	1 V	5 V
1	10 V	10 V

Table E.1. Behavior of the logarithmic function generator

It should be noted that the LOG112 has a rather limited bandwidth, so its application in simulations involving high time scale factors should be given careful consideration.

A sine/cosine generator

In some cases, such as the inverted pendulum simulation which was used to train a machine learning system, basic harmonic functions such as $\sin(\varphi)$ and $\cos(\varphi)$ cannot be generated using the quadrature generator described in section 6.33, figure 6.118, because computing element variations will cause $\dot{\varphi}$ and the derived functions $\sin(\varphi)$ and $\cos(\varphi)$ to drift apart over time.

If the argument φ can be guaranteed to be in an interval like $[-1, 1]$, functions like $\sin(\varphi)$, etc., can be generated using special function generators based on polygonal approximation. The AD639[228] is a rare but useful integrated circuit which generates basic trigonometric functions; this was used to build the dedicated function generator shown in figure F.1.

Both potentiometers R2 and R6 have to be adjusted so that an input voltage in the range of $[-2\pi, 2\pi]$ rad $\hateq [-1, 1] \hateq [-10, 10]$ V is mapped to the interval $[-7.2, 7.2]$ V.

---

228 See [Analog Devices].

F A sine/cosine generator

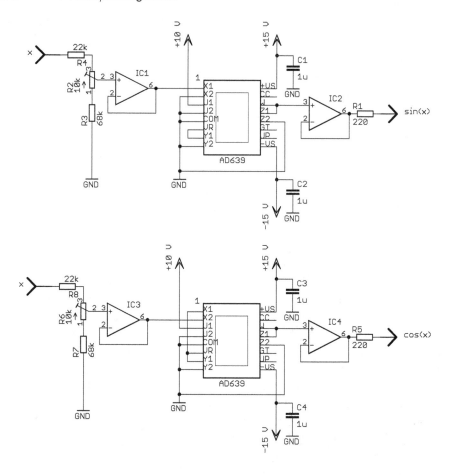

Fig. F.1. Sine/cosine generator based on the AD639 universal trigonometric function converter

A simple joystick interface

Since an analog computer is ideally suited for all kinds of dynamic systems simulations, a joystick interface is a really versatile peripheral device as it makes "man in the loop" simulations possible. This appendix describes a simple adapter circuit which allows a two-channel analog joystick, as commonly used for controlling models, to be used with an analog computer.

The joystick used here is a two-channel joystick from an old model remote control. It consists of two 4k7 (4700 Ω) potentiometers – one for the x- and one for the y-direction. A simple handheld enclosure was built, as shown on the left in figure G.2. This also holds a push button which can be used to control an electronic switch or a comparator in the analog computer.

The circuit shown in figure G.1 is pretty straightforward: First, the two reference voltages of ± 10 V are buffered by the operational amplifiers IC1A and IC1B. R1 and R2 are the joystick's potentiometers while R1* and R2* are used to set the origin of the joystick (putting the joystick into its middle position with respect to x and y should yield a value of 0 on each channel).

Since the potentiometers used in a typical joystick allow for a 270 degree travel and as the joystick only allows for a much smaller deflection, the wiper will never hit its end positions. Accordingly, the output signals x and y must be amplified to satisfy 10 V $\leq x, y \leq +10$ V. This is done by the two operational amplifiers IC1C and IC1D. The capacitors C1 and C2 suppress any unwanted oscillations which otherwise may occur. The slight non-linearity due to the load resistance of 10 k on the wipers of the joystick potentiometers is negligible in this application, so no additional buffer stage is required.

The switch S1 which is also mounted in the joystick's enclosure can be seen on the right side of the schematic in figure G.1. As long as it is open the output is tied to +10 V by means of a 100k resistor. Closing the switch will yield a negative

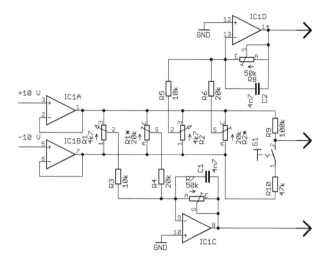

Fig. G.1. Joystick adapter circuit

Fig. G.2. Joystick mounted in enclosure and prototype adapter circuit

value so that either a comparator or an electronic switch (in the case of an Analog Paradigm Model-1 analog computer) can be controlled by this switch.

The right half of figure G.2 shows the prototype circuit built on a standard 160×100 mm^2 breadboard. The joystick is plugged into the 9-pin SUB-D connector visible on the upper left of the circuit board. The 2 mm sockets yield the output signals (x and y as well as the signal from the push button switch).

The Analog Paradigm bus system

The Analog Paradigm model-1 analog computer features a bus system that makes it easy to extend the system with additional modules for special operations such as the logarithm function generator described in section E. All computing modules are based on standard *Eurocards* (160 mm×100 mm) using two-row (A/C) DIN 41612 connectors. Table H.1 shows the signals available on this bus system.

When designing new hardware for the Analog Paradigm Model-1 analog computer, the following constraints should be taken into account:

- Signals denoted by an overline are active-low logic signals.
- Analog and digital ground (AGND/DGND) should never be connected to each other on a module as this would introduce a ground loop and deteriorate system performance.
- When $\overline{\text{POTSET}}$ is active, i.e., low, the computer is set to potentiometer setting mode. The integrators are halted and the inputs of all potentiometers are connected to the positive reference voltage. Using the readout feature described below, the value set on each potentiometer can then be displayed on a DVM.
- The $\overline{\text{OVERLOAD}}$ line can be driven by a computing module to signal an overload condition to the control unit. The driver must be of the open collector type. (A standard TTL-driver can be used but must be decoupled by a diode.)
- The signals MODEIC and MODEOP control the operation of the integrators of the overall system.
- To read the output value of a computing element its address must be placed onto the address lines A15–A0 and the $\overline{\text{READ}}$ line must be driven low. The element selected in this way will then connect its output to the READOUT line by means of an electronic switch (or a relay). In addition to this, each

A	Pin no.	C
+15 V	1	+15 V
−15 V	2	−15 V
AGND	3	AGND
+10 V	4	+10 V
−10 V	4	−10 V
AGND	6	AGND
READOUT	7	READOUT
AGND	8	AGND
AGND	9	AGND
$\overline{\text{POTSET}}$	10	$\overline{\text{OVERLOAD}}$
MODEOP	11	MODEIC
RS4	12	RS3
RS2	13	RS1
A15	14	A14
A13	15	A12
A11	16	A10
A9	17	A8
A7	18	A6
A5	19	A4
A3	20	A2
A1	21	A0
	22	
	23	
RS0	24	$\overline{\text{CSELECT}}$
$\overline{\text{WRITE}}$	25	$\overline{\text{READ}}$
D7	26	D6
D5	27	D4
D3	28	D2
D1	29	D0
SCL	30	SCA
+5 V	31	+5 V
DGND	32	DGND

Table H.1. The Analog Paradigm Model-1 bus system

computing element can drive the data lines (D7–D0) to show its module type which can then be displayed on a readout unit.

- The 16 address bits consist of a rack number (A15–A12 – this will be typically 0000), a chassis number (A11-A8 – from bottom to top in a rack), a slot number (A7–A4 – from left to right), and an element number (A3–A0). Each card can hold up to 16 computing elements.
- The lines labeled RS, SCA, SCL, and $\overline{\text{WRITE}}$ are reserved for future use.

HyCon commands

The Model-1 hybrid controller, HC, is connected by means of a serial line USB interface to a digital computer which sends appropriate commands to control the operation of the analog computer. Using a serial line terminal program it is possible to control the analog computer by typing in these commands:[229]

A: Enable halt-on-overflow. When enabled, any overload condition will immediately HALT the analog computer so that the element(s) causing this condition can be identified by their respective overload indicators.

a: Disable halt-on-overflow.

B: Enable external halt – a trigger signal on the HALT-input of the hybrid controller will place the analog computer into HALT mode regardless of its current mode of operation.

b: Disable external halt.

c: This command expects a six-digit decimal number which specifies the OP-time for repetitive or single-run operation in milliseconds. If values less than 10^5 milliseconds are required, which is typically the case, the number has to be padded with zeros on the left.

C: This command also expects a six digit decimal number specifying the time for setting initial conditions in milliseconds.

d: Clear the digital output port specified by an address in the range of 0 to 7.

D: Like d but set the digital output port specified by a single-digit address following the command.

---

[229] It should be noted that these commands should not be followed by a carriage return (<CR>) and/or line feed (<LF>) control character. The hybrid controller firmware parses its input data in a bytewise fashion and is not record oriented. Spurious <CR>/<LF> or whitespace characters will result in error messages as they are regarded as illegal commands.

e: Issuing this command starts repetitive operation. The analog computer will cycle through the modes IC and OP with the respective times set by the c- and C-commands until it is halted.

E: This command starts a single-run cycle, i. e., the analog computer runs through the sequence IC, OP, HALT with the IC- and OP-times as specified above.

f: If a readout-group has been defined with the G-command, issuing an f will read all of the elements in that group sequentially and return their values.

F: This command, too, starts a single-run cycle. The main difference to the E-command is that a message will be returned at the end of the cycle, allowing the analog and digital computer to be synchronized.

g: To read the value of a single computing element, the g-command, followed by a four digit hexadecimal address, is used.[230] The hybrid controller will return the ID of the computing element as well as its current value.

G: This command expects a semicolon-separated list of four digit hexadecimal addresses which must be terminated by a single dot to let the hybrid controller know that no more entries follow.

h: Set the analog computer into HALT-mode.

i: Set the analog computer into IC-mode.

l: Return all values sampled during the last OP cycle if a readout group had been defined previously.

o: Set the analog computer to OP-mode.

P: This command for setting a digital potentiometer expects the four digit hexadecimal address of the module containing the potentiometer to be set, followed by a two digit address of the potentiometer, followed by a four digit decimal (!) value in the range of 0000 to 1023.[231] To set the digital potentiometer number three on the module with the base address 0x0090 to 1/2, the command P0090030511 would be required.

q: This prints out the current digital potentiometer setting.

R: Read all eight digital input ports and return their respective values.

s: Print out status information.

S: Set the analog computer into POTSET-mode. This is useful for setting manual potentiometers – see section 2.5.

t: During an OP-phase this command will print the elapsed time since the OP-mode started.

x: Reset the hybrid controller.

X: Configure a crossbar switch module.[232]

?: Print help.

---

[230] The addressing scheme is described in appendix H.
[231] The value 1023 corresponds to a coefficient of 1.
[232] At the time of writing this, X is only an experimental feature.

The following example shows how the hybrid controller can be controlled by manually issuing appropriate commands.[233] First, the OP- and IC-times are set to 500 and 100 milliseconds respectively as echoed by the controller (lines 1 to 4). Then a repetitive run is started by issuing an e-command which is eventually terminated by forcing the analog computer into HALT-mode (line 6). Then the digital output with address 0 is set to high before another repetitive run is started and later terminated by the h-command.

The command in line 10 reads the value from the computing element with the hexadecimal address 0160. The hybrid controller returns 0.0061 1, the first part being the value read while the second number is the identification number of the computing element being read.

Line 12 defines a readout group consisting of two computing elements with the respective hexadecimal addresses 0160 and 0161. Issuing an s-command returns complete status information of the hybrid controller also containing the addresses of all elements in the readout group. The f-command in line 16 reads all elements of this group and returns their values as -0.0004;-0.0002.

```
─────────── example.yml ───────────
1   c000500
2   T_OP=500
3   C000100
4   T_IC=100
5   e
6   h
7   D0
8   e
9   h
10  g0160
11  0.0061 1
12  G0160;0161.
13  s
14  STATE=NORM,MODE=IC,EXTH=DIS,OVLH=DIS,IC-time=100,OP-time=500,
15  MYADDR=90,RO-GROUP=352;353
16  f
17  -0.0004;-0.0002
─────────── example.yml ───────────
```

---

233 The line feeds after each command have been inserted for better readability. The hybrid controller does not expect line feed control characters after commands.

Bibliography

[ADLER 1968] H. ADLER, *Elektronische Analogrechner*, VEB Deutscher Verlag der Wissenschaften, Berlin, 1968

[ALBRECHT 1968] PETER ALBRECHT, "Über die Behandlung linearer partieller Differentialgleichungen auf einem hybriden Rechensystem", Vortrag, gehalten zum 4. Internationalen Kongress INTERKAMA, 1968, in *Elektronische Rechenanlagen*, Heft 6, 1968, pp. 280–285

[AMDAHL 1967] GENE M. AMDAHL, "Validity of the Single Processor Approach to Achieving Large-Scale Computing Capabilities", in *AFIPS Conference Proceedings* (30): 483-485, 1967

[AMELING 1963] WALTER AMELING, *Aufbau und Wirkungsweise elektronischer Analogrechner*, Friedrich Vieweg & Sohn, Braunschweig, 1963

[AMMON 1966] W. AMMON, *Schaltungen der Analogrechentechnik*, R. Oldenbourg, 1966

[Analog Devices] Analog Devices, *Universal Trigonometric Function Converter AD639*

[BABERUXKI 2022] NICK BABERUXKI, *Quantum Mechanical Two-Body Problem with Gaussian Potential*, Analog Computer Applications, https://analogparadigm.com/downloads/alpaca_36.pdf, retrieved December 2nd, 2022

[BATTIN 1999] RICHARD H. BATTIN, *An introduction to the mathematics and methods of astrodynamics*, American Institute of Aeronautics and Astronautics, Inc., Reston, Virginia, 1999

[BECK et al. 1958] ROBERT M. BECK, MAX PALEVSKY, "The DDA", in *Instruments and Automation*, November 1958, pp. 1836–1837

[BELKHIR et al. 2018] LOTFI BELKHIR, AHMED ELMELIGI, „Assessing ICT global emissions footprint: Trends to 2040 & recommendations", in *Journal of Cleaner Production* 117, Elsevier, 2018, pp. 448–463

[BEKEY et al. 1968] GEORGE A. BEKEY, WALTER J. KARPLUS, *Hybrid Computation*, John Wiley & Sons, Inc., 1968

[BENEKING 1971] HEINZ BENEKING, *Praxis des Elektronischen Rauschens*, Bibliographisches Institut, 1971

[BLOCH et al. 2010] ANTHONY M. BLOCH, ALBERTO G. ROJO, "Sorting: The Gauss Thermostat, the Toda Lattice and Double Bracket Equations", in [Hu et al. 2010, p. 35 et seq.]

[BORCHARDT] INGE BORCHARDT, *Kepler und die Atomphysik – Der Beschuss eines Atomkerns mit Alphateilchen auf einem Tischanalogrechner*, Demonstrationsbeispiel 3, AEG-Telefunken, ADB 003 0570

[BOWMAN 1958] FRANK BOWMAN, *Introduction to Bessel Functions*, Dover Publications, Inc., 1958

[BRACHER et al. 2021] JOHANNES BRACHER, DANIEL WOLFRAM, TILMANN GNEITING, MELANIE SCHIENLE, "Vorhersagen sind schwer, vor allem die Zukunft betreffend: Kurzzeitprognosen in der Pandemie", in *Mitteilungen der Deutschen Mathematiker-Vereinigung*, 2021, 29, 4, pp. 186–190

[BROCK 2019] DAVID C. BROCK, "The Shocking Truth Behind Arnold Nordsieck's Differential Analyzer", in *IEEE Spectrum*, 30. Nov. 2017, https://spectrum.ieee.org/tech-history/dawn-of-electronics/the-shocking-truth-behind-arnold-nordsiecks-differential-analyzer, retrieved September 24th, 2019

[BROCKETT 1991] R. W. BROCKETT, "Dynamical Systems That Sort Lists, Diagonalize Matrices, and Solve Linear Programming Problems", in *Linear Algebra and its Applications*, 146 (1991), Elsevier Science Publishing Co., Inc., pp. 79–91

[BRONSTEIN et al. 1989] I. N. BRONSTEIN, K. A. SEMENDJAJEW, *Taschenbuch der Mathematik*, 24. Auflage, Verlag Harri Deutsch, Thun und Frankfurt/Main, 1989

[BRYANT et al. 1962] LAWRENCE T. BRYANT, MARION J. JANICKE, LOUIS C. JUST, ALAN L. WINIECKI, *Engineering Applications of Analog Computers*, Argonne National Laboratory, ANL-6319, October 1962

[BYWATER 1973] R. E. H. BYWATER, *The Systems Organization of Incremental Computers*, PhD thesis, University of Surrey, 1973

[CARLSON et al. 1967] ALAN CARLSON, GEORGE HANNAUER, THOMAS CAREY, PETER J. HOLSBERG (eds.), *Handbook of Analog Computation*, 2nd edition, Electronic Associates, Inc., Princeton, New Jersey, 1967

[CARSLAW et al. 1941] H. S. CARSLAW, J. C. JAEGER, *Operational Methods in Applied Mathematics*, Oxford at the Clarendon Press, 1941

[CHANG 2017] TAI-PING CHANG, "Chaotic Motion in Forced Duffing System Subject to Linear and Nonlinear Damping", in *Hindawi – Mathematical Problems in Engineering*, Volume 2017, Article ID 3769870, 15

[CHROMATIC 2015] CHROMATIC, *Modern Perl*, Onyx Neon, Inc., 2015, http://www.onyxneon.com/books/modern_perl/index.html, retrieved October 3rd, 2019

[COPE 2017] GREG COPE, *The Aizawa Attractor*, http://www.algosome.com/articles/aizawa-attractor-chaos.html, retrieved December 15th, 2018

[CUNNINGHAM 1954] W. J. CUNNINGHAM, *Time-Delay Networks for an Analog Computer*, Office of Naval Reserach, Nonr-433(00), Report No. 6, August 1, 1954

[DOETSCH 1970] GUSTAV DOETSCH, *Einführung in Theorie und Anwendung der Laplace-Transformation*, Birkhäuser Verlag Basel und Stuttgart, 1970

[DUFFING 1918] GEORG DUFFING, "Erzwungene Schwingungen bei veränderlicher Eigenfrequenz und ihre technische Bedeutung", Sammlung Vieweg, Heft 41/42, Braunschweig 1918

[DUFFY 1994] DEAN G. DUFFY, *Transform Methods for Solving Partial Differential Equations*, CRC Press, 1994

[EAI PACE 231R] EAI, *PACE 231R analog computer*, Electronic Associates, Inc., Long Branch, New Jersey, Bulletin No. AC 6007

[EAI 1964] EAI, *Solution of MATHIEU's Equation on the Analog Computer*, Applications Reference Library, Application Study: 7.4.4a, 1964

[EAI 1.3.2 1964] EAI, *Continuous Data Analysis with Analog Computers Using Statistical and Regression Techniques*, EAI Applications Reference Library 1.3.2a, 1964, http://bitsavers.org/pdf/eai/applicationsLibrary/1.3.2a_Continuous_Data_Analysis_with_Analog_Computers_Using_Statistical_and_Regression_Techniques_1964.pdf

[FANO 1950] R. M. FANO, "Short-Time Autocorrelation Functions and Power Spectra", in *The Journal of the Acoustical Society of America*, Volume 22, Number 5, pp. 546–550

[FENECH et al. 1973] HENRI FENECH, JAY BANE, "A nuclear reactor analog simulation for undergraduate nuclear engineering education", in *Simulation*, Volume 20, Issue 4, April 1974, pp. 127–135

[FIFER 1961] STANLEY FIFER, *Analogue Computation – Theory, Techniques and Applications*, Vol. III, McGraw-Hill Book Company, Inc., 1961

[FISCHER 2022] THOMAS FISCHER, *First Steps – THE ANALOG THING*, https://the-analog-thing.org/THAT_First_Steps.pdf, retrieved December 14th, 2022

[FÖLLINGER et al. 2021] OTTO FÖLLINGER, MATHIAS KLUWE, *Laplace-, Fourier- und z-Transformation*, VDE Verlag GmbH, 11. Auflage, 2021

[FORBES 1957] GEORGES F. FORBES, *Digital Differential Analyzers*, Fourth Edition, 1957

[FORBES 1972] GEORGE FORBES, "The simulation of partial differential equations on the digital differential analyzer", in ACM '72 Proceedings of the ACM annual conference, Volume 2, pp. 860–866

[FREETH 2008] TONY FREETH, *The Antikythera Mechanism – Decoding an Ancient Greek Mystery*, Whipple Museum of the History of Science, University of Cambridge, 2008

[FREETH 2010] TONY FREETH, "Die Entschlüsselung eines antiken Computers", in *Spektrum der Wissenschaft*, April 2010, p. 62 et seq.

[GELENBE et al. 2015] EROL GELENBE, YVES CASEAU, "The Impact of Information Technology on Energy Consumption and Carbon Emissions", in *Ubiquity*, Volume 2015, June 2015, Article No. 1

[GILES 1976] CHRISTOPHER BRUCE GILES, *The Development of a Hybrid Simulator for Power System Control Investigations*, PhD Thesis, Imperial College London, 1976

[GILLILAND 1967] MAXWELL C. GILLILAND, *Handbook of Analog Computation (Including Application of Digital Control Logic)*, SYSTRON-DONNER CORP., 1967

[GILOI et al. 1963] WOLFGANG GILOI, RUDOLF LAUBER, *Analogrechnen*, Springer-Verlag, Berlin/Göttingen/Heidelberg, 1963

[GOLDBERG et al. 1954] E. A. GOLDBERG et al., *Stabilized Direct Current Amplifier*, United States Patent 2684999, July 27, 1954

[GOLDMAN 1965] MARK W. GOLDMAN, "Design of a High Speed DDA", in AFIPS '65 (Fall, part I) Proceedings of the November 20–December 1, 1965, fall joint computer conference, part I, pp. 929–949

[HAUSNER 1971] ARTHUR HAUSNER, *Analog and Analog/Hybrid Computer Programming*, Prentice-Hall, Inc., 1971

[HAVIL 2019] JULIAN HAVIL, *Curves for the Mathematically Curious – an Anthology of the Unpredictable, Historical, Beautiful, and Romantic*, Princeton University Press, 2019

[HEDDEGHEM et al. 2014] WARD VAN HEDDEGHEM, SOFIE LAMBERT, BART LANOO, DIDIER COLLE, MARIO PICKAVET, PIET DEMEESTER, "Trends in worldwide ICT electricity consumption from 2007 to 2012", in *Computer Communications*, Volume 50, 1 September 2014, pp. 64–76

[HENZE 2018] NORBERT HENZE, *Irrfahrten – Faszination der Random Walks*, Springer Spektrum, 2. Auflage, 2018

[HINDMARSH et al. 1982] J. L. HINDMARSH, R. M. ROSE, "A model of the nerve impulse using two first-order differential eequations", in *Nature*, Vol. 296, 11th March 1982, pp. 162–164

[HINDMARSH et al. 1984] J. L. HINDMARSH, R. M. ROSE, "A model of neuronal bursting using three coupled first order differential equations", in *Prov. R. Soc. Lond.*, B 221, 87–102 (1984)

[Höhler 1988] G. Höhler, "Schwingungen II", in *Computer – Theoretikum und Praktikum für Physiker*, Fachinformationszentrum Karlsruhe, 1988, pp. 5–73

[Höltgen et al. 2016] Stefan Höltgen, Jan Claas van Treeck (Hrsg.), *Time to Play – Zeit und Computerspiel*, Verlag Werner Hülsbusch, 2016

[Holzer et al. 2020] Mirko Holzer, Bernd Ulmann, "Hybrid computer approach to train a machine learning system", in *Alternative Computing*, World Scientific Publishers, to appear in 2020

[Hort 1910] Wilhelm Hort, *Technische Schwingungslehre*, Verlag von Julius Springer, Berlin, 1910

[Howe 1961] R. M. Howe, *Design Fundamentals of Analog Computer Components*, D. van Nostrand Company, Inc., 1961

[Howe 1961] Robert Howe, "Analog Techniques", in *Instruments & Control Systems*, Vol. 34, 1961, p. 1482 et seq.

[Hu et al. 2010] Xiaoming Hu, Ulf Jonsson, Bo Wahlberg, Bijoy K. Gosh (eds.), *Three Decades of Progress in Control Sciences*, Springer-Verlag Berlin Heidelberg, 2010

[Huang et al. 2017] Yipeng Huang, Ning Guo, Mingoo Seok, Yannis Tsividis, Simha Sethumadhavan, "Analog Computing in a Modern Context: A Linear Algebra Accelerator Case Study" in *IEEE MICRO*, May/June 2017, pp. 30–38

[Huskey et al. 1962] Harry D. Huskey, Granino A. Korn, *Computer Handbook*, McGraw-Hill Book Company, Inc., 1962

[Jackson 1960] Albert S. Jackson, *Analog Computation*, McGraw-Hill Book Company, Inc., 1960

[Johnson 1956] Clarence L. Johnson, *Analog Computer Techniques*, McGraw-Hill Book Company, Inc., 1956

[Jones 2018] Nicola Jones, „The Information Factories – Data centres are chewing up vast amounts of energy – so researchers are trying to make them more efficient", in *Nature*, Springer, Vol. 561, pp. 163–166

[Jung 2006] Walter G. Jung, *Op Amp Applications Handbook*, Analog Devices, Inc., 2006

[Karplus et al. 1958] Walter J. Karplus, Walter W. Soroka, *Analog Methods – Computation and Simulation*, McGraw-Hill Book Company, Inc., 1958

[Kennedy 1962] Jerome D. Kennedy Sr., "Representation of Time Delays", in [Huskey et al. 1962, pp. 6-3–6-16]

[Kiers et al. 2003] Ken Kiers, Tim Klein, Jeff Kolb, Steve Price, "Chaos in a nonlinear analog computer", in *International Journal of Bifurcation and Chaos*, Vol. 14, No. 8 (2004), pp. 2867–2873

[Klein et al. 1957] Martin L. Klein, Frank K. Williams, Harry C. Morgan, "Digital Differential Analyzers", in *Instruments and Automation*, June 1957, pp. 1105–1109

[KNORRE 1971] WOLFGANG A. KNORRE, *Analog computer in Biologie und Medizin – Einführung in die dynamische Analyse biologischer Systeme*, VEB Gustav Fischer Verlag, Jena, 1971

[KORN et al. 1964] GRANINO A. KORN, THERESA M. KORN, *Electronic Analog and Hybrid Computers*, McGraw-Hill Book Company, Inc., 1964

[KORN 1966] GRANINO A. KORN, *Random-Process Simulation and Measurement*, McGraw-Hill Book Company, 1966

[KORN et al. 1972] GRANINO A. KORN, THERESA M. KORN, *Electronic Analog and Hybrid Computers*, McGraw-Hill Book Company, 1972

[KORSCH et al. 2008] HANS JÜRGEN KORSCH, HANS-JÖRG JODL, TIMO HARTMANN, "The Duffing Oscillator", in *Chaos – A Program Collection for the PC*, Springer, 2008, pp. 157–184

[KORTE 1964] EDWARD L. KORTE, "Stable Wide-Frequency Oscillator", in *Simulation*, 1964, 2, p. 31

[KOVACH et al. 1962] L. D. KOVACH, H. F. MEISSINGER, "Solution of Algebraic Equations, Linear Programming, and Parameter Optimization", in [HUSKEY et al. 1962, pp. 5-133–5-154]

[KUEHN 2015] CHRISTIAN KUEHN, *Multiple Time Scale Dynamics*, Springer, 2015

[LANCHESTER 1908] FREDERICK WILLIAM LANCHESTER, *Aerial Flight: Aerodonetics*, London, Constable, 1908

[LANGFORD 1984] WILLIAM FINLAY LANGFORD, "Numerical Studies of Torus Bifurcations", in *International Series of Numerical Mathematics*, Vol. 70, 1984 Birkhäuser Verlag Basel, pp. 285–295

[LEVI 2012] MARK LEVI, *The Mathematical Mechanic – Using Physical Reasoning to Solve Problems*, Princeton University Press, eighth printing, 2012

[LIAO 2013] SHIJUN LIAO, *Chaotic motion of three-body problem – an origin of macroscopic randomness of the universe*, arXiv: 1304.2089v1, 8. Apr. 2013

[LORENZ 1963] EDWARD NORTON LORENZ, "Deterministic Nonperiodic Flow", in *Journal of the Atmospheric Sciences*, Volume 20, pp. 130–141, March 1963

[LORENZ 1984] EDWARD N. LORENZ, *Irregularity: a fundamental property of the atmosphere*, Tellus (1984), 36A, pp. 98–110

[LOTZ AAB] H. LOTZ, *Darstellung von Tragflügeln und ihren Stromlinien mit einem Analogrechner*, AEG Telefunken, AAB 011 1069

[LOTZ ASD] H. LOTZ, *Perspektivische Darstellung von Rechenergebnissen mit Hilfe eines Analogrechners*, AEG Telefunken, ASD 051 0469

[MAHRENHOLTZ 1968] OSKAR MAHRENHOLTZ, *Analogrechnen in Maschinenbau und Mechanik*, BI Hochschultaschenbücher, 1968

[MASSEN 1977] ROBERT MASSEN, *Stochastische Rechentechnik – Eine Einführung in die Informationsverarbeitung mit zufälligen Pulsfolgen*, Carl Hanser Verlag, München, Wien, 1977

[MCLACHLAN 1961] NORMAN WILLIAM MCLACHLAN, *Bessel Functions for Engineers*, Oxfort at the Clarendon press, reprint of the second edition, 1961

[MORRISON 1962] E. MORRISON, "Nuclear-Reactor Simulation", in [HUSKEY et al. 1962, pp. 5-87–5-93]

[MUNZ 2014] PHIL MUNZ, "When Zombies Attack! Alternate Ending", in [SMITH? 2014, pp. 45–55]

[Meccano 1934] N. N., "Machine Solves Mathematical Problems – A Wonderful Meccano Mechanism", in *Meccano Magazine*, Vol. XIX, No. 6, June, 1934, pp. 442–444

[MOULTON 1923] FOREST RAY MOULTON, *An Introduction to Celestial Mechanics*, The Macmillan Company, Second Revised Edition, 1923

[MÜLLER 1986] HERIBERT MÜLLER, "Simulation und Lösung physikalischer Probleme mit dem Analogrechner", in *Praxis der Naturwissenschaften, Physik*, Aulis Verlag, Heft 3/35, 15. April 1986, pp. 21–25

[NORDSIECK 1953] ARNOLD NORDSIECK, *The Nordsieck computer*, AIEE-IRE '53 (Wester) Proceesings of the February 4–6, 1953, estern computer conference, pp. 227–231

[OTTERMAN 1960] JOSEPH OTTERMAN, "The Properties and Methods for Computation of Exponentially-Mapped-Past Statistical Variables", in *IRE Transactions on Automatic Control*, Volume: AC-5, Issue: 1, 1. Jan. 1960, pp. 11–17, DOI: 10.1109/TAC.1960.6429289

[Panasonic 3007] Panasonic, *MN3007 – 1024-stage low noise BBD*

[Panasonic 3101] Panasonic, *MN3101 – clock generator/driver CMOS LSI for BBD*

[PRESS et al. 2007] WILLIAM H. PRESS, SAUL A. TEUKOLSKY, WILLIAM T. VETTERLING, BRIAN P. FLANNERY, *Numerical Recipes*, Cambridge University Press, Third Edition, 2007

[REUTTER et al. 1968] FRITZ REUTTER, JOSEF NEUKIRCHEN, DIETMAR SOMMER, *Herstellung konformer Abbildungen mit Hilfe des Analogrechners – Praktische Behandlung der Umströmung zweifach zusammenhängender Gebiete*, Westdeutscher Verlag, Köln und Opladen, 1968

[RHEINBOLDT 2009] WERNER C. RHEINBOLDT, *Classical Iterative Methods for Linear Systems*, https://www-m2.ma.tum.de/foswiki/pub/M2/Allgemeines/SemWs09/lin-iter.pdf, retrieved February 1st, 2023

[RÖPKE et al. 1969] HORST RÖPKE, JÜRGEN RIEMANN, *Analogcomputer in Chemie und Biologie – Eine Einführung*, Springer-Verlag, 1969

[RÖSSLER 1976] OTTO E. RÖSSLER, "An Equation for Continuous Chaos", in *Physics Letters*, Volume 57 A, number 5, pp. 397–398

[RUBY 1996] LAWRENCE RUBY, "Applications of the Mathieu equation", in *American Journal of Physics*, 64 (1), January 1996, pp. 39–44

[SACHS 2016] JASON SACHS, *Padé Delay is Okay Today*, https://www.embeddedrelated.com/showarticle/927.php, March 1, 2017, retrieved April 13th, 2018

[SAWHNEY et al. 2020] ROHAN SAWHNEY, KEENAN CRANE, "Monte Carlo Geometry Processing: A Grid-Free Approach to PDE-Based Methods on Volumetric Domains", *ACM Trans. Graph.*, Vol 38, No. 4, Article 1, July 2020, pp. 1-1–1-18

[SAWHNEY et al. 2022] ROHAN SAWHNEY, DARIO SEYB, WOJCIECH JAROSZ, KEENAN CRANE, "Grid-Free Monte Carlo for PDEs with Spatially Varying Coefficients", *ACM Trans. Graph.*, Vol. 41, No. 4, Article 53, July 2022, pp. 53-1–53-17

[SCHABACK 2020] ROBERT SCHABACK, "On COVID-19 Modelling", in *Jahresbericht der Deutschen Mathematiker-Vereinigung*, 2020, Jun 29, pp. 1–39

[SCHÖNEFELD 1977] REINHOLD SCHÖNEFELD, *Hybrid-Simulation*, Akademie-Verlag Berlin, 1977

[SCHWANKNER 1980] ROBERT SCHWANKNER, *Radiochemie-Praktikum*, UTB, Schöningh, 1980

[SCHWARZ 1971] WOLFGANG SCHWARZ, *Analogprogrammierung – Theorie und Praxis des Programmierens für Analogrechner*, VEB Fachbuchverlag Leipzig, 1. Ed., 1971

[SENSICLE 1968] ALLAN SENSICLE, *Introduction to Control Theory for Engineers*, Blackie & Son Limited, 1968

[SHILEIKO 1964] A. V. SHILEIKO, *Digital Differential Analyzers*, Pergamon Press, The Macmillan Company, New York, 1964

[SIMANCA et al. 2002] SANTIAGO R. SIMANCA, SCOTT SUTHERLAND, *Notes for MAT 331 – Mathematical Problem Solving with Computers*, The University at Stony Brook, https://www.math.stonybrook.edu/~scott/Book331/331book.pdf

[SMALL 2001] JAMES S. SMALL, *The Analogue Alternative – The Electronic Analogue Computer in Britain and the USA, 1930–1975*, Routledge, London and New York, 2001

[SMITH? 2014] ROBERT SMITH?, *Mathematical Modelling of Zombies*, University of Ottawa Press, 2014

[SOROKA 1962] WALTER W. SOROKA, "Mechanical Analog Computers", in [HUSKEY et al. 1962, pp. 8-2–8-16]

[SPROTT 2010] JULIEN CLINTON SPROTT, *Elegant Chaos – Algebraically Simple Chaotic Flows*, World Scientific Publishing Co. Pte. Ltd, 2010

[STUBBS et al. 1954] G. S. STUBBS, C. H. SINGLE, *Transport Delay Simulation Circuits*, Westinghouse, Atomic Power Division, 1954

[SUTTON et al. 2018] RICHARD S. SUTTON, ANDREW G. BARTO, *Reinforcement Learning – An Introduction*, second edition, The MIT Press, Cambridge, Massachusetts, London, England, 2018

[SYDOW 1964] ACHIM SYDOW, *Programmierungstechnik für elektronische Analogrechner*, VEB Verlag Technik, Berlin, 1964

[Telefunken/Ball] Telefunken, *Demonstrationsbeispiel Nr. 5: Ball im Kasten*,

[Telefunken/Particle] Telefunken, *Demonstrationsbeispiel: Elektrisch geladenes Teilchen im Magnetfeld*, AEG-Telefunken, ADB 007

[Telefunken 1963] N. N., *Anwendungsbeispiele für Analogrechner – Wärmeleitung*, Telefunken, 15. Oktober 1963

[THOMAS et al. 1969] CHARLES H. THOMAS, RONALD A. HEDIN, "Switching Surges on Transmission Lines Studied by Differential Analyzer Simulation", in *IEEE Transactions on Power Apparatur and Systems*, Vol. PAS-88, No. 5, May 1969

[TRUITT et al. 1960] THOS. D. TRUITT, A. E. ROGERS, *Basics of Analog Computers*, John F. Rider Publisher, Inc., New York, December 1960

[TYROR et al. 1970] J. G. TYROR, R. I. VAUGHAN, *An Introduction to the Neutron Kinetics of Nuclear Power Reactors*, Pergamon Press, 1970

[ULMANN 2010] BERND ULMANN, *Analogrechner: Wunderwerke der Technik – Grundlagen, Geschichte und Anwendung*, Oldenbourg, 2010

[ULMANN 2023] BERND ULMANN, *Analog Computing*, 2nd edition, DeGryuter, 2023

[ULMANN 2016] BERND ULMANN, "Man in the Loop – Zeitaspekte in analogen Simulationen und Spielen", in [HÖLTGEN et al. 2016], pp. 95–119

[ULMANN et al. 2019] BERND ULMANN, DIRK KILLAT, "Solving systems of linear equations on analog computers", in *2019 Kleinheubach Conference*, Miltenberg, Germany, 2019, pp. 1–4, IEEE Xplore, 04. November 2019

[VAN DER POL et al. 1928] BALTHASAR VAN DER POL, J. VAN DER MARK, "The Heartbeat considered as a Relaxation Oscillation, and an Electrical Model of the Heart", in *Phil. Mag.* 7, 1928, pp. 763–775

[VAN DER POL 1987] BALTHASAR VAN DER POL, *Operational Calculus Based on the Two-Sided Laplace Integral*, Chelsea Publishing Company, New York, 3rd edition, 1987

[VOLYNSKII et al. 1965] B. A. VOLYNSKII, V. YE. BUKHMAN, *Analogues for the Solution of Boundary-Value Problems*, Pergamon Press, 1965

[WAGNER 1972] MANFRED WAGNER, *Analogrechner in der Verfahrenstechnik*, Dr. ALFRED HÜTHIG Verlag Heidelberg, 1972

[WEEDY et al. 1998] BIRRON M. WEEDY, BRIAN J. CORY, *Electric Power Systems*, 4th edition, John Wiley & Sons Ltd., 1998

[WEINBERG et al. 1958] ALVIN M. WEINBERG, EUGENE P. WIGNER, *The Physical Theory of Neutron Chain Reactors*, The University of Chicago Press, 1958

[WILLERS 1943] FRIEDRICH ADOLF WILLERS, *Mathematische Instrumente*, Verlag von R. Oldenbourg, München und Berlin 1943

[WIDDER 2010] DAVID VERNON WIDDER, *The Laplace Transform*, Dover Publications, Inc., 2010

[WINKLER 1961] HELMUT WINKLER, *Elektronische Analogieanlagen*, Akademie-Verlag Berlin, 1961

[Yavetz 1995] Ido Yavetz, *From Obscurity to Enigma – The Work of Oliver Heaviside, 1872–1889*, Birkhäuser Verlag, Basel, Boston, Berlin, 1995

[Yokogawa] *Analog Computers Series 3300*, Yokogawa Electric Works, Ltd., Catalog No. YEW 3300A

[Zhan et al. 2016] Lusa Zhan, Yipeng Huang, "Analog Sorting – Theory and evaluation", https://yipenghuang.com/wp-content/uploads/2017/03/analog_sorting.pdf, retrieved December 13th, 2022

Index

$T_{1/2}$, 51
Δ, 183
k_0, 19
∇, 183
$\propto$, 50
$\perp$, 17

A4 rocket, 7
absolute value, 82
acceleration, 49
ADC, 32, 93
AI, 1, 245
airfoil, 176
AIZAWA attractor, 156
algebraic loop, 190
all-pass filter, 98
AMDAHL' law, 244
AMDAHL's law, 8
amplifier
    high gain, 16
    open, 16, 76, 177
    operational, 12
amplitude stabilization, 58
analog
    computer, 1
    program, 47

analog-digital-converter, 32, 93
analogon, 1
analogue, 1
analogy, 1
    direct, 2
    indirect, 2
analyzer
    differential, 4
angle-perserving mapping, 179
Antikythera, 3
Antikythera mechanism, 3
Arduino®, 93, 261
artificial intelligence, 1, 245
attractor
    AIZAWA, 156
    CHUA, 151
    LORENZ, 148, 149
    RÖSSLER, 146
AVOGADRO's constant, 54

backlash, 84
ball
    bouncing, 141
ballistic trajectory, 130
banana plug, 39
bang-bang, 86

BBD, 95
BESSEL functions, 125
BLACK, HAROLD STEPHEN, 13
block diagram, 254
bottleneck, von NEUMANN, 244
bouncing ball, 141
break-point, 30
BROMWICH integral, 252
BROMWICH, THOMAS JOHN L'ANSON, 252
BROWNIAN motion, 238
bucket brigade, 89
bucket brigade device, 95
buffered potentiometer, 27
BUSH, VANNEVAR, 4

capacitor wheel, 92
capture cross section, 145
carriage return, 277
celestial mechanics, 138
chaos
    nonlinear, 155
charged particle, 131
chemical kinetics, 110
CHUA attractor, 151
CHUA diode, 151
CHUA, LEON ONG, 151
circuit, 2
clothoid, 165
CMP4, 35
coefficient, 25
    diffusiuon, 238
    drift, 238
comparator, 35
compensation voltmeter, 27
computer
    analog, 1
    hybrid, 1, 219
    stochastic, 2
conductivity
    thermal, 183
conformal mapping, 176, 179

Control Transformer, 7
converter
    analog/digital, 93
    digital/analog, 93
CR, 277
critical damping, 68
critical section, 8
CU, 42

DAC, 32, 93
damped pendulum, 117
damping
    critical, 68
    subcritical, 68
DDA, 2
dead zone, 84
decay, 49
    rate, 50
decay rate, 51
definite
    positive, 191
del operator, 127
delay, 91
delayed neutron, 214
delayed unit step function, 248
delta function, 249
density, 183
DEQ, 47
DEREK DE SOLLA PRICE, 3
derivative
    time, 89
differential analyzer, 4
differential equation, 47
    ordinary, 49
    partial, 17, 49, 180
    stiff, 8
diffusion coefficient, 238
diffusivity
    thermal, 182
Digital Differential Analyzer, 2
digital voltmeter, 27
digital-analog-converter, 32, 93

diode, 30
    ideal, 80
DIRAC, PAUL ADRIEN MAURICE, 249
direct analogy, 2
division, 76
Double Scroll attractor, 151
drag, 130
drift coefficient, 238
DUFFING oscillator, 161
DVM, 27, 30, 54, 275

EAI, 19, 31
eigenfrequency, 66
Electronic Associates Inc., 19, 31
EMP, 103
equation
    MATHIEU's, 119
    VOLTERRA-LOTKA-, 145
    heat, 182
    wave, 49
EULER spiral, 165
Eurocard, 275
excitation function, 117
exponentially mapped past, 103

feedback, 51
FET, 95
FEYNMAN-KAC formula, 238
field effect transistor, 95
field programmable gate array, 1
filter
    all-pass, 98
    low pass, 78
fire control system, 7
flow, 176
forcing function, 117
four quadrant multiplier, 34
FPGA, 1
free elements, 24
free potentiometer, 25, 59
frequency

eigen-, 66
natural, 66
friction-wheel, 4
function
    delayed unit step, 248
    delta, 249
    excitation, 117
    forcing, 117
    generator, 30
    harmonic, 56
    inverse, 74
    ramp, 249
    special, 73
    step, 248
    transfer, 96
    unit step, 248

glider, 171
GND, 13
ground, 13

half-life, 51
halt, 18, 22
HALT, 18, 22
harmonic function, 56
heat equation, 182
heat transfer, 180
HEAVISIDE, OLIVER, 248
Hermitian matrix, 191
high gain amplifier, 16
high performance computing, 1, 8
HINDMARSH-ROSE model, 168
HOELZER, HELMUT, 7
HPC, 1, 8
human-in-the-loop, 193
hybrid computer, 1, 32, 219
hysteresis, 85

IC, 18, 21, 33
ideal diode, 80
impedance converter, 27
indirect analogy, 2

individual mode control, 89
initial condition, 17, 18, 21
input
    table, 4
    weight, 12
input/output, 36
INT4, 20
integrated circuit, 33
integrator, 17
    leaky, 105
inverse function, 74
inverter, 23
isotope, 49

jerk, 49
jolt, 49
JOUKOWSKY airfoil, 176
JOUKOWSKY, NIKOLAY
    YEGOROVICH, 176, 179
joystick, 273

KELVIN, 4
KELVIN feedback technique, 4, 51, 98
kinetics
    chemical, 110
KIRCHHOFF, GUSTAV, 14
KUTTA, MARTIN WILHELM, 179
KUTTA-JOUKOWSKY transform, 179

LAPLACE delay operator, 96
LAPLACE operator, 183
LAPLACE transform, 97, 247
LAPLACE, PIERRE SIMON, 247
leaky integrator, 105
LF, 277
limiter, 83
line feed, 277
linear equations
    systems of, 187
loaded potentiometer, 26
loop
    algebraic, 190
LORENZ attractor, 148, 149
LORENZ, EDWARD NORTON, 148, 149
low pass filter, 78

machine
    learning, 245
    time, 6, 20, 55
    unit, 11
    variable, 12, 54
MACNEE, A. B., 7
magnetic field, 131
MAGNUS effect, 176
MAGNUS, GUSTAV HEINRICH, 176
mapping
    angle-preserving, 179
    conformal, 176, 179
mass-spring-damper, 65
mathematical pendulum, 62
MATHIEU's equation, 119
matrix, 189
    Hermitian, 191
    transposed, 191
maximum, 87
MDS2, 76
mean value, 79
methane chlorination, 113
MICHAELIS-MENTEN kinetics, 113
min/max holding, 87
minimum, 87
MLT8, 34
mode
    control
        individual, 89
    halt, 18, 22
    HALT, 18, 22
    IC, 18, 21
    initial condition, 18
    OP, 18, 22
    operate, 18, 22
model, 2

modulation, 86
mol, 54
multiplier, 33
    four quadrant, 34
    quarter-square, 33, 34
    two quadrant, 34
music, 210

nabla operator, 127
natural frequency, 66
negative feedback, 13
neuron, 168
neutron
    delayed, 214
neutron kinetics, 214
NOSÉ-HOOVER oscillator, 157

ODE, 49
OP, 18, 22
opamp, 12
open amplifier, 16, 76, 177
open-loop gain, 13
operate, 18, 22
operational amplifier, 12
operator
    del, 127
    nabla, 127
order, 49
    reaction, 110
ordinary differential equation, 49
oscillation
    phugoid, 171
Oslo differential analyzer, 6
overdamped, 68
overload, 11

PADÉ approximation, 98
partial differential equation, 17, 49, 180
particle
    charged, 131
particular solution, 57, 77

passband, 78
patch
    field, 1
    panel, 1
PDE, 49
pendulum, 62
    damped, 117
phase
    diagram, 118
    space plot, 118, 125
PHILBRICK, GEORGE A., 7
phugoid oscillation, 171
PID-controller, 236
planimeter, 4
plot
    phase space, 118, 125
plotter, 6, 23, 36
plug
    banana, 39
POISSON equation, 238
polynomials, 77
Polyphemus, 7
positive definite, 191
potential flow, 177
potentiometer, 25
    buffered, 27
    free, 25, 59
    loaded, 26
    servo, 39
    setting, 27
    unloaded, 26
POTSET, 27
powers, 77
predator-prey-system, 145
problem time, 20, 55
problem variable, 12, 54
program
    analog, 47
PT8, 28
Python, 222, 231

quadrature generator, 57

quarter-square multiplier, 33, 34

Rössler attractor, 146
radioactiv decay, 49
RAM, 93
ramp function, 249
random access memory, 93
random walk, 238
RC, 256
reaction speed, 110
reinforcement learning, 230
relaxation parameter, 236
repetitive operation, 7, 23, 36
reset, 18
residue theorem, 252
resolver, 197
Richardson method, 236
RLC-circuit, 70
Rössler, Otto, 146
rotating spiral, 163
Rutherford-scattering, 135

sample and hold, 88, 93, 94
scaling, 11, 54, 55
    time, 12, 20, 55
    variable, 12
SEIR model, 114
Selsyn, 7
servo-potentiometer, 39
single run, 23, 36
SJ, 14
slide rule, 4
smooth sorting, 216
solution
    particular, 57, 77
sorting
    smooth, 216
source term, 238
Space Technology Laboratories, 219
special functions, 73
specific heat capacity, 183
speed
    reaction, 110
spiral
    rotating, 163
SQ_M model, 160
square
    root, 75
    wave, 80
stabilization
    amplitude, 58
step function, 248
Stieltjes integral, 73
stiff differential equation, 8
stochastic computer, 2
Stubbs-Single approximation, 101
subcritical damping, 68
substitution technique, 51
SUM8, 17
summer, 12
summing junction, 14
sweep, 61
switch, 35
synchro, 7
system
    predator-prey-, 145
systems of linear equations, 187

Taylor series, 30, 65
Telefunken, 31
THAT, 43
THE ANALOG THING, 43
thermal
    conductivity, 183
    diffusivity, 182
Thomson, William, 4
time
    delay, 91
    machine, 6, 20, 55
    problem, 20, 55
    scale factor, 18, 19
    scaling, 12, 20, 55
time derivative, 89
trajectory, 130

transfer function, 96, 103, 254
transfer system, 96
transistor
    field effect, 95
Transmitter, 7
transposed matrix, 191
triangle
    signal, 80
    wave, 80
two quadrant multiplier, 34

underdamped, 68
unit impulse, 249
unit step function, 248
unloaded potentiometer, 26

V2 rocket, 7
VAN DER POL, 122
variable
    machine, 12, 54
    problem, 12, 54
    scaling, 12
velocity, 49

    potential, 177
voltage divider, 25
VOLTERRA-LOTKA-equations, 145
voltmeter
    compensation, 27
    digital, 27
von NEUMANN bottleneck, 244

wave
    equation, 49
    square, 80
    triangle, 80
weight, 12
WIENER process, 238
wobbulator, 61

XIBNC, 25
XID, 24
XIR, 24

Z-diode, 24
ZENER-diode, 24
Zombie apocalypse, 144

Printed in the USA
CPSIA information can be obtained
at www.ICGtesting.com
JSHW050228170424
61266JS00009B/181